T0220081

Cambridge Elements ☰

Elements in Decision Theory and Philosophy
edited by
Martin Peterson
Texas A&M University

MONEY-PUMP ARGUMENTS

Johan E. Gustafsson
University of York
University of Gothenburg
Institute for Futures Studies

CAMBRIDGE
UNIVERSITY PRESS

Shaftesbury Road, Cambridge CB2 8EA, United Kingdom

One Liberty Plaza, 20th Floor, New York, NY 10006, USA

477 Williamstown Road, Port Melbourne, VIC 3207, Australia

314–321, 3rd Floor, Plot 3, Splendor Forum, Jasola District Centre,
New Delhi – 110025, India

103 Penang Road, #05–06/07, Visioncrest Commercial, Singapore 238467

Cambridge University Press is part of Cambridge University Press & Assessment, a department of the University of Cambridge.

We share the University's mission to contribute to society through the pursuit of education, learning and research at the highest international levels of excellence.

www.cambridge.org
Information on this title: www.cambridge.org/9781108718950
DOI: 10.1017/9781108754750

First published 2022

A catalogue record for this publication is available from the British Library.

ISBN 978-1-108-71895-0 Paperback
ISSN 2517-4827 (online)
ISSN 2517-4819 (print)

Cambridge University Press & Assessment has no responsibility for the persistence or accuracy of URLs for external or third-party internet websites referred to in this publication and does not guarantee that any content on such websites is, or will remain, accurate or appropriate.

Money-Pump Arguments

Elements in Decision Theory and Philosophy

DOI: 10.1017/9781108754750
First published online: September 2022

Johan E. Gustafsson
University of York
University of Gothenburg
Institute for Futures Studies

Author for correspondence: Johan E. Gustafsson,
johan.eric.gustafsson@gmail.com

Abstract: Suppose that you prefer A to B, B to C, and C to A. Your preferences violate Expected Utility Theory by being cyclic. Money-pump arguments offer a way to show that such violations are irrational. Suppose that you start with A. Then you should be willing to trade A for C and then C for B. But then, once you have B, you are offered a trade back to A for a small cost. Since you prefer A to B, you pay the small sum to trade from B to A. But now you have been turned into a money pump. You are back to the alternative you started with but with less money. This Element shows how each of the axioms of Expected Utility Theory can be defended by money-pump arguments of this kind. This title is also available as Open Access on Cambridge Core.

Keywords: money pumps, dynamic choice, rationality, cyclicity, Expected Utility Theory

ISBNs: 9781108718950 (PB), 9781108754750 (OC)
ISSNs: 2517-4827 (online), 2517-4819 (print)

Contents

1 Money-Pump Arguments

It's 1955. You've been offered a full professorship with a salary of $5,000. The dean calls you to his office to go over the final details. On his desk lie three contracts – labelled *A*, *B*, *C*.

"As you know, your current offer (contract *A*) is a full professorship at $5,000," the dean says, handing you the contract. "Yet... a little birdie told me you prefer an assistant professorship at $6,000 (contract *C*) since it pays a lot more. Potentially, I could offer you the assistant professorship. Potentially."

The dean rubs his thumbs over his index and middle fingers – the gesture for money. Message received. You slip him $20.

"It's a pleasure to offer you the assistant professorship," the dean says, handing you contract *C* in exchange for *A*. "Even so, I've been informed you prefer an associate professorship at $5,500 (contract *B*) since it's more prestigious for just a little bit less money. Might that be worth something to you?"

The dean rubs his fingers. All right. You slip him another $20.

"It's, ahem, a pleasure to offer the associate professorship," the dean says, handing you contract *B* in exchange for *C*. "Still, I've heard you prefer your first offer (contract *A*) since it's still more prestigious for only slightly less money."

The dean rubs his fingers: You slip him another $20.

"It is a real pleasure to *re*offer you the full professorship," the dean says, handing you contract *A* in exchange for *B*. "Well deserved."

Once more, you got your first offer, but now you've lost $60 to the dean – who, by the way, is only getting started. The dean nods towards contract *C*, rubbing his fingers.[1]

You've been reduced to a money pump![2] You've become the dean's private cash dispenser. A victim of your own mind, you were brought to ruin by the structure of your preferences. You preferred *A* to *B*, *B* to *C*, and *C* to *A*. This cycle of preferences left you open to blatant exploitation. So such cyclic preferences must, it seems, be irrational.

Arguments of this kind let us demonstrate that some alleged requirement of rationality really is a requirement of rationality. A *money-pump argument* for some alleged requirement of rationality consists of an argument that otherwise rational agents who violate the requirement would in some possible situation

---

[1] Davidson, McKinsey, and Suppes (1955, pp. 145–6) introduced this first money-pump example, crediting Norman Dalkey.

[2] The term 'money pump' dates back to Edwards, Lindman, and Phillips 1965, p. 273.

end up paying for something they could have kept for free even though they knew in advance what decision problem they were facing.

We will investigate whether there are compelling money-pump arguments that rational preferences conform to Expected Utility Theory, which is a structural requirement on preferences over prospects. Let a *final outcome* be a description of the world that captures everything that the agent cares about.[3] Let a *prospect* be a probability distribution over all potential final outcomes.[4] (And let a *sure prospect* be a prospect with a single possible final outcome.[5]) Expected Utility Theory, then, is the theory that prospects are preferred in accordance with an expected-utility function:[6]

> *Expected Utility Theory*   Let Ω be the set of possible final outcomes and $p_X(o)$ be the probability of outcome o in prospect X. Then there is a real-valued function u such that, for all prospects X and Y, it holds that X is at least as preferred as Y if and only if
>
> $$\sum_{o \in \Omega} u(o)p_X(o) \geq \sum_{o \in \Omega} u(o)p_Y(o).$$

Rather than this general form, we will be concerned with Expected Utility Theory restricted to prospects with *finite support* – that is, prospects with a finite number of final outcomes with positive probability. Given this restriction, Expected Utility Theory is entailed by the following basic axioms:[7]

- Completeness (Section 3)
- Transitivity (Section 4.1)

---

[3] Arrow 1965, p. 12.

[4] Marschak 1950, p. 114.

[5] Marschak 1950, p. 114.

[6] Bernoulli 1738, p. 177; 1954, p. 24, von Neumann and Morgenstern 1944, pp. 24–5, and Jensen 1967, pp. 172–3. For a precursor to Expected Utility Theory which deals with expectations of value rather than utility, see Arnauld and Nicole 1662, p. 467; 1996, pp. 273–4 and Pascal 1670, pp. 56–7; S680/L418; 2004, pp. 212–13.

[7] For proof, see Jensen 1967, pp. 172–82, Fishburn 1970, pp. 111–5, and Hammond 1998, pp. 152–64. For the proof, we also need to assume the standard algebra of combining prospects. See von Neumann and Morgenstern 1944, pp. 26–7 and Jensen 1967, p. 170. To derive a more general form of Expected Utility Theory which also holds for prospects with *countably infinite support* (that is, prospects with a countably infinite number of final outcomes with positive probability), we need to supplement the four basic axioms with the following requirement of rationality:

> *Equiprobable Weak-Preference Dominance*   If there are prospects $X_1, X_2, \ldots$ and $Y_1, Y_2, \ldots$ and probabilities $p_1, p_2, \ldots$ such that the probabilities sum to 1, and, for all $i = 1, 2, \ldots$, it holds that $X_i \succeq Y_i$ and $p_i > 0$, then the prospect of, for all $i = 1, 2, \ldots$, probability p_i of X_i is at least as preferred as the prospect of, for all $i = 1, 2, \ldots$, probability p_i of Y_i.

See Blackwell and Girshick 1954, p. 105 and Hammond 1998, pp. 189–90.

- The strong strict-preference version of Independence (Section 5.3)
- Continuity (Section 6)

Our main task will be to show, with the help of money-pump arguments, that these axioms are requirements of rationality. For the first three axioms, we will find compelling money-pump arguments.[8] But, for Continuity, we will only be able to find an argument that is almost a money-pump argument.

Money-pump arguments are often dismissed due to a number of influential objections – for example: (i) that you could rationally avoid being money pumped if you use foresight, (ii) that you could rationally avoid being money pumped if you are resolute and stick to a plan, and (iii) that money-pump arguments prove too much, because, in some cases with infinite series of trade offers, even agents who conform to Expected Utility Theory are exploitable.

We will rebut these and other objections. While foresight blocks the standard version of the money-pump argument, there are other versions that work for agents who use foresight (Section 2.1). Once the resolute approach is spelled out in detail, the problems with escaping money pumps by being resolute become apparent (Section 7). And agents who conform to Expected Utility Theory avoid exploitation even in cases with infinite series of trade offers, as long as they use foresight (Section 8).

We won't start with the axioms of Expected Utility Theory, however. Rather, we'll start with the most discussed money-pump argument: the argument that rational preferences are acyclic.[9]

2 Acyclicity

2.1 Three-Step Acyclicity

Consider having a cup of coffee with one, two, or three lumps of sugar. Suppose that you can't taste the difference between a cup with one lump and a cup with two lumps. Nor can you taste the difference between a cup with two lumps and a cup with three lumps. And, when you can't taste any difference, you prefer

---

[8] My ordering of the basic axioms isn't arbitrary. The money-pump argument for Completeness does not rely on the other basic axioms. The argument for Transitivity relies on Completeness. The argument for the strong strict-preference version of Independence relies on Completeness and Transitivity. And the argument for Continuity relies on all the other basic axioms.

[9] There are also money-pump arguments for other requirements of rationality. Notably, there are money-pump arguments that rational credences satisfy the laws of probability. (See Ramsey 1931, p. 182.) These arguments are known as Dutch-book arguments. (See Lehman 1955, p. 251.) For an overview, see Pettigrew 2020.

having less sugar (to keep your intake down). Still, you can taste the difference between a cup with one lump and a cup with three lumps – and, due to your sweet tooth, you prefer the latter.[10]

Let A, B, and C be the sure prospects of having a cup with one, two, and three lumps of sugar respectively. You prefer A to B, B to C, and C to A. Let '$X > Y$' denote that X is (strictly) preferred to Y.[11] Then we can state your preferences as follows:[12]

(1) $A > B > C > A$.

Your preferences are cyclic. More specifically, your preferences violate the following requirement:[13]

> *Three-Step Acyclicity* If $X > Y > Z$, then it is not the case that $Z > X$.

All violations of Three-Step Acyclicity have the same form as the preferences in (1). So, to show that Three-Step Acyclicity is a requirement of rationality, all we need to show is that preferences of the kind in (1) are irrational.

The standard version of the money-pump argument runs as follows.[14] Suppose that you start off with A. An exploiter offers you a trade from A to C. Since you prefer C to A, you accept this offer. Then, after this first trade, you

---

[10] This is a variation of an example in Dummett 1984, p. 34. Both examples, however, have the same structure of indiscernibility as an example in Armstrong 1939, p. 457n1 and one in Luce 1956, p. 179 (which also involves coffee with varying amounts of sugar). Ng (1977, p. 52) presents a similar example based on indiscernibility in three dimensions, where no dimension is more important than the others. Another source of cyclic preferences is majority rule. Suppose that roughly one third of the people have the preferences $A > B > C$, that roughly one third of the people have the preferences $B > C > A$, and that roughly one third of the people have the preferences $C > A > B$. Now, suppose that you are a dedicated democrat who prefers an option over another if it is preferred to the other option by a majority of the people. Then you have the preferences $A > B > C > A$. See Condorcet 1785, p. lxi; 1976, p. 54; 1994, p. 124, Dodgson 1876, pp. 8–12, and Black 1948, pp. 32–3. May (1954, p. 6) presents a similar example, where you assess options on three criteria, and you prefer an option if it is superior on at least two of these criteria. Then, in the same way as in the majority-rule example, you end up with cyclic preferences. A further example of cyclic preferences is the Mere-Addition Paradox in population ethics (see note 66). Finally, you might prefer one die over another if it is more likely to win than the other, which leads to cyclic preferences – given a special kind of dice. See Gardner 1970, p. 110.

[11] Debreu 1959, p. 8.

[12] We adopt the convention that chains of relations with overlapping relata can be contracted. For example, '$X > Y$ and $Y > Z$' can be written as '$X > Y > Z$'. See De Morgan 1851, p. 104 and Fishburn 1991, p. 116.

[13] Samuelson 1947, p. 151 and Sen 1977, p. 62. For the statement of principles, we adopt the convention that free variables are implicitly universally quantified.

[14] Edwards et al. 1965, p. 273 and Pratt et al. 1965, ch. 2, p. 10. Davidson et al. (1955, p. 146) present a similar case, but it's not clear that they meant to show that the agent ends up paying for what they could have kept for free. See Gustafsson 2013, p. 462 for the interpretation that Davidson et al. only intended to illustrate that the agent with the preferences in (1) would regret any choice from the option set $\{A, B, C\}$, which may be a sign of irrationality in itself. For this alternative argument, see Tullock 1964, p. 403.

are offered a second trade from C to B. Since you prefer B to C, you also accept this second trade. Finally, after the second trade, you are offered a third trade from B to A^-, where A^- is just like A except that you have less money and

(2) $A > A^- > B$.

We can let the payment be monetary to fit the exploitation framing, but the main point is that A^- is less preferred than A by being certainly inferior with respect to some dimension you care about and the same in other respects.[15] Let a *souring* of a prospect X be a prospect that is just like X except that it is certainly inferior in a dimension the agent cares about. (And let a *sweetening* of a prospect X be a prospect that is just like X except that it is certainly superior in a dimension the agent cares about.)

That there is an A^- like this follows from (1) by the following requirement of rationality:[16]

> *Unidimensional Continuity of Preference* If $X > Y$, then there is a prospect X^- such that (i) X^- is a souring of X and (ii) $X > X^- > Y$.

The idea is that, if you (strictly) prefer X to Y, you must prefer X with some margin. So there should be some, perhaps minimal, amount you're willing to pay to get X rather than Y.[17] This is what blocks trivial responses to the

---

[15] There's no special relationship between money and rationality. If you don't prefer having more money, there's nothing irrational about freely losing money. See Gustafsson 2013, pp. 462–4.

[16] Hansson 1993, p. 478. Note that this principle differs from Continuity, which we won't assume. (We will *defend* Continuity in Section 6.) Unidimensional Continuity of Preferences, in contrast to Continuity, does not cover continuity of probability for preferences over prospects. (This also applies to Weak Insensitivity to Souring in Section 3 and Unidimensional Continuity of Dispreference in Section 4.2.)

[17] It may seem strange that a requirement of rationality would have any implications for what things exist. Note, however, that the things that are mentioned in the requirement are only taken to exist as possibilia, not as concrete things. The same worry about existential implications also applies to Continuity (see Section 6). A potential solution is to replace the existential quantifier with a 'some' quantifier that lacks existential implications. See Priest 2005, pp. 11–14. Another worry is that Unidimensional Continuity of Preference prohibits rational agents from caring exclusively about discrete things in final outcomes (such as money in the smallest possible denomination or the smallest noticeable differences of pleasure). To deal with this worry, we could rely on the following probabilistic variant:

> *Stochastic Unidimensional Continuity of Preference* If $X > Y$, then there is a prospect X^- and some probability p such that (i) X^- is a souring of X, (ii) $0 \leq p < 1$, and (iii) $XpX^- > Y$.

Here, XpX^- is a prospect consisting in a lottery between X and X^- such that X occurs with probability p and X^- occurs with probability $1 - p$. If we rely on this alternative principle, the exploitation would only be probabilistic. If a money-pump set-up makes you choose XpX^- over X, then the exploiter only gets your money with some probability but they still risk nothing. These adjustments could also be made for the other principles we'll assume about the existence of sourings and sweetenings.

money-pump argument where the agent avoids exploitation by preferring not to pay for anything.[18]

Now, since you prefer A^- to B, you also accept the third trade. So you end up with A^- (that is, you pay for A) when you could have kept A for free.

In this example, you followed an approach known as *myopic choice* – that is, you assessed each choice in isolation, as if it were the only choice you would ever make.[19] We distinguish myopic choice from *naive choice*, which is to (i) consider the prospects of all available plans and assess which of these prospects are choice-worthy in a choice between all of them and (ii) choose in accordance with a plan to end up with one of these choice-worthy prospects – without taking into consideration whether you would later depart from that plan.[20]

Do you avoid exploitation if you follow naive choice rather than myopic choice? To follow naive choice in this case, you need to first consider the prospects of the available plans (that is, A, A^-, B, and C) and assess which of these prospects are choice-worthy. But how do you choose among three or more prospects if you have cyclic preferences over those prospects? Consider the Maximization Rule:[21]

> *The Maximization Rule*　It is rationally permitted to choose a prospect X if and only if there is no feasible prospect Y such that $Y > X$.

Given your cyclic preferences, you can't maximize in a choice between A, A^-, B, and C, since each prospect is less preferred than another prospect. We avoid this problem with the following alternative rule:[22]

> *The Uncovered-Choice Rule*　It is rationally permitted to choose a prospect X if and only if there is no feasible prospect Y such that $Y > X$ and, for all feasible prospects Z, it holds that $Y > Z$ if $X > Z$.

So far, we haven't made any assumptions about your preference between A^- and C. Yet, since you prefer C to A, you may, plausibly, also prefer C to a souring of A.[23] So we could plausibly suppose

---

[18]　See, for instance, Ahmed 2014, p. 29n14.

[19]　Following Dow 1984, p. 96 and McClennen 1990, pp. 11–12.

[20]　Following Pollak 1968, pp. 202–3 and Hammond 1976, p. 162.

[21]　Uzawa 1956, p. 37.

[22]　Schwartz 1990, p. 21; see also Miller 1980, pp. 72–4. Schwartz (1970, pp. 105–6) also proposed the following alternative rule:

> *The Generalized Optimal-Choice Rule*　Faced with a choice from a set of prospects U, it is rationally permitted to choose a prospect if and only if it is in any subset V of U such that (i), for all X in V and for all Y that are in U but not in V, it is not the case that $Y > X$ and (ii) no non-empty proper subset of V satisfies (i).

This rule, however, permits the choice of A^- from $\{A, A^-, B, C\}$, given (1), (2), and (3). See Fishburn 1977, p. 478.

[23]　This inference is an instance of Unidimensional IP-Transitivity (see Section 4.1).

(3) $C > A^-$.

Then, given the Uncovered-Choice Rule, the only prospects you would be rationally permitted to choose from A, A^-, B, and C are A, B, and C. Note that A^- is ruled out because A is preferred to A^- and to every option that A^- is preferred to.

Even so, does the naive approach avoid exploitation in this case? It does not. Naive choice combined with the Uncovered-Choice Rule still allows you to accept the first two trades, since you regard B as choice-worthy. So you are rationally permitted to choose in accordance with the plan to accept the first two trades to get B. Then, when you face the third offer of trading from B to A^-, the Uncovered-Choice Rule allows you to choose A^- (because A is no longer the prospect of some available plan). So adopting naive choice does not save you from exploitation.

The standard version of the money-pump argument isn't very convincing, however. In order for a money-pump argument to be compelling, the agent must know in advance what decision problem they face – that is, they must know the whole exploitation set-up in advance.[24] The exploiter must not rely on any knowledge about what will happen that is unavailable to the agent, because being exploited by someone who knows more than you need not be a sign of irrationality.[25] But, if you know the whole exploitation set-up in advance, you can use foresight to see that some of the trades aren't in your interest.[26] To see how, consider the decision tree of the standard money-pump set-up in Figure 1, which we can call the *Standard Money Pump*.

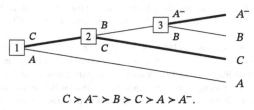

$$C > A^- > B > C > A > A^-.$$

Figure 1 The Standard Money Pump

Here, the three squares represent choice nodes corresponding to the three consecutive trade offers. You accept a trade by going up at the corresponding

---

[24] Anand 1993a, pp. 61–2; 1993b, p. 341.
[25] Schwartz 1986, pp. 129–30. Briggs (2010, p. 13) suggests that, rather than whether the exploiter knows anything the agent does not, what's crucial is that the exploiter's behaviour matches the agent's expectations.
[26] Schick 1986, pp. 117–18.

choice node, and you turn the trade down by going down.[27] The item on the upper left of each square is what you get if you accept the trade, and the item on the lower left is what you keep if you turn the trade down (that is, what you give up if you accept the trade). The preferences stated below the decision tree are the agent's preferences, which are held constant throughout.[28] So, in this case and in the other cases we'll consider, agents do not revise their preferences during the decision problem.[29]

If you have foresight, you can use backward induction. To use *backward induction* is to predict what you would choose at later choice nodes and to take those predictions into account when you choose at earlier nodes.[30] First,

---

[27] Following the convention in Rabinowicz 2008, p. 152.

[28] Hence all your time-slices throughout the decision problem have the same preferences. This breaks Hedden's (2015a, p. 430; 2015b, pp. 113–17) companions-in-guilt defence against money-pump arguments. Hedden argues that, in a Prisoner's Dilemma, rational players also end up with an outcome that is dominated for everyone by another outcome they could have achieved together. A *Prisoner's Dilemma* is a game with two players: Row, who has the preferences $B > A > A^- > C$, and Column, who has the preferences $C > A > A^- > B$. Row and Column each face a choice between cooperation and defection:

		Column	
		Cooperate	Defect
Row	Cooperate	A	C
	Defect	B	A^-

By dominance reasoning, holding the other player's move fixed, both players will defect. So they end up with A^- when they could have achieved A together. (See Luce and Raiffa 1957, pp. 94–5 and Tucker 1980, p. 101.) The problem for the companions-in-guilt defence is that, in a Prisoner's Dilemma, it's crucial that the players have different preferences (if they didn't, there would be no dilemma). But, in the decision problems we will consider, all your time-slices have the same preferences. So one would expect that they should be able to avoid this kind of diachronic tragedy. Likewise, since your time-slices act in sequence with full knowledge of what the earlier time-slices have chosen, they shouldn't have any problems coordinating.

[29] Seidenfeld 1988, p. 275. Anand (1993a, p. 62; 1993b, p. 341), Hansson (1993, p. 484), and Nozick (1993, p. 160) object that agents must be allowed to revise their preferences. While there seems to be nothing irrational about revising one's preferences, this isn't germane to our discussion. The money-pump argument targets preferences with a certain structure and tries to show that having those preferences may turn you into a money pump. The fact that you could avoid becoming a money pump by adopting some *other* preferences does not vindicate your original preferences. If you have to give up your cyclic preferences to avoid becoming a money pump, then the money-pump argument has succeeded. For a discussion of the Revision Approach to resolute choice, see Section 7. Hansson (1993, p. 484) suggests that it's the introduction of new options which brings about the preference change; but, for the decision problems we will discuss, we assume that the agent knows the whole decision problem in advance.

[30] von Neumann and Morgenstern 1944, pp. 116–17, Strotz 1955–6, p. 173, Raiffa and Schlaifer 1961, pp. 7–8. For an early use of backward induction, see Cayley 1875, p. 237. (Backward induction is also known as sophisticated choice; see Pollak 1968, p. 203 and Hammond 1976, p. 162.) A common complaint is that backward induction prescribes permanent defection in

consider the trade at node 3. At this node, you have a choice between A^- and B. And, since you prefer A^- to B, you would accept the trade to A^- at node 3. (The choices that are prescribed by backward induction are marked by the thicker lines in the decision tree.) This assumes that the only thing that should guide your choice at a node is your preference between the still feasible options; we accept the following principle:[31]

> *Decision-Tree Separability* The rational status of the options at a choice node does not depend on other parts of the decision tree than those that can be reached from that node.

In what follows, we'll take Decision-Tree Separability for granted (until Section 7, where we take on challenges to this kind of separability).

Using backward induction at node 2, we take into account the prediction that A^- would be chosen at node 3. Given this prediction, accepting the trade at node 2 effectively results in your final holding being A^- whereas turning it down results in your final holding being C. Since you prefer C to A^-, you would turn down the trade at node 2.

And, taking this prediction into account at node 1, we find that accepting the trade at node 1 effectively results in your final holding being C whereas turning it down results in your final holding being A. Since you prefer C to A, you accept the trade at node 1. Hence, using backward induction, you will end up with C after accepting the first trade and then turning the second. So you avoid paying for something you could have kept for free. And so the standard money-pump argument is blocked.[32]

Nevertheless, we can revise the exploitation set-up so that it works against people who use backward induction.

One way to do so is to repeat the trade offers in case they are rejected but with no more than three trade offers in total, as in the decision problem in Figure 2, the *Money Pump with Repeated Offers*.[33]

---

finite iterations of Prisoner's Dilemma (described in note 28); see Selten 1978, pp. 136–8. But, as argued by Sobel (1993, pp. 130–1), there seems to be nothing wrong with the backward-induction reasoning in such cases. Even so, he claims, that conclusion has little relevance for typical games with imperfectly rational players.

[31] McClennen 1988, p. 522; 1990, p. 122. Note that this is a weaker principle than Hammond's (1988a, p. 508) Normal-Form Consequentialism (see note 60). Some versions of resolute choice (notably the Counter-Preferential Approach) violate Decision-Tree Separability. We'll discuss them in Section 7.

[32] McClennen 1990, p. 166 and Rabinowicz 1995, p. 593.

[33] Rabinowicz 2000a, p. 141.

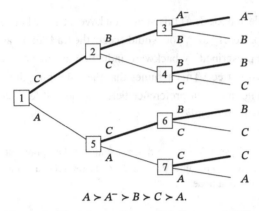

$$A \succ A^- \succ B \succ C \succ A.$$

Figure 2 The Money Pump with Repeated Offers

At any of the final nodes (that is, nodes 3, 4, 6, and 7), you would accept the trades you are offered. In other words, you would go up at each of these nodes.

Using backward induction at nodes 2 and 5, you take into account the prediction that you would accept each of the final trades. So the choice at node 2 is effectively between A^- (accepting the trade) and B (turning it down). Since you prefer A^- to B, you would accept the trade at node 2 (that is, you would go up to node 3). Likewise, the choice at node 5 is effectively between B (accepting the trade) and C (turning it down). Since you prefer B to C, you would accept the trade at node 5 (that is, you would go to node 6).

Using backward induction at node 1, you take all of these predictions into account. So the choice at node 1 is effectively between A^- (accepting the trade) and B (turning it down). Since you prefer A^- to B, you accept this first trade. So you go up to node 2 and then up to node 3, where you finally choose A^-. And then you end up with A^- even though you could have kept A for free.

Still, this argument relies upon a contentious form of backward induction. Specifically, one may challenge our assumption that you would choose rationally and retain your trust in your future rationality even at choice nodes that could only be reached by irrational choices.[34] To see the problem, suppose that you predict not only rational choices but also some irrational choices in the Money Pump with Repeated Offers. We mark these predicted choices at nodes that would follow irrational choices by thick dashed lines in Figure 3.

---

[34] Binmore 1987, pp. 196–200.

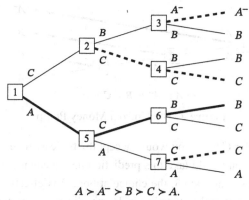

$$A > A^- > B > C > A.$$

Figure 3 The Money Pump with Repeated Offers
(blocked by predicted irrationality)

Taking these predictions into account at node 1, the choice at that node is effectively between *C* (accepting the trade) and *B* (turning it down). Since you prefer *B* to *C*, it's rationally required that you turn down the trade at node 1. And the move to node 2 is irrational. So the predicted irrational choices would only be made at nodes that follow irrational choices. Hence, in order to rule out these predictions, we must assume that you would choose rationally and retain your trust in your future rationality even at nodes that follow irrational choices. But it's implausible that you would be rationally required to retain your trust in your future rationality at nodes where you've already made irrational choices.[35]

This worry about backward induction does not apply to decision problems that are BI-terminating. A decision problem is *BI-terminating* if and only if the choices that are prescribed by backward induction are terminal, that is, the prescribed choices are not followed by any potential further choice nodes.[36] To defend the prescriptions of backward induction in BI-terminating decision problems, we only need to assume that, at nodes that can be reached without making any irrational choices, you retain (i) your rationality and (ii) your trust in your rationality at nodes that can be reached without making any irrational choices. While the money pumps we've considered so far are not BI-terminating, the *Upfront Money Pump* in Figure 4 is BI-terminating.[37]

---

[35] Gustafsson and Rabinowicz 2020, p. 582.

[36] Rabinowicz 1998, p. 101.

[37] Gustafsson and Rabinowicz 2020, p. 583. For a similar construction in voting theory, see Moulin 1983, pp. 96–7.

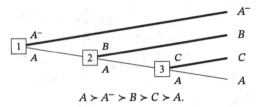

$$A \succ A^- \succ B \succ C \succ A.$$

Figure 4 The Upfront Money Pump

Since you prefer C to A, you would accept the trade at node 3. Using backward induction, you would take this prediction into account at node 2. Given the predicted choice at node 3, the choice at node 2 is effectively between B (accepting the trade) and C (turning it down). Since you prefer B to C, you would accept the trade at node 2. And, taking this prediction into account at node 1, the initial choice is effectively between A^- (accepting the trade) and B (turning it down). Since you prefer A^- to B, you accept the trade at node 1 and end up with A^- even though you could have kept A for free.

This backward-induction argument assumes that you would also make rational choices and retain your trust in your future rationality at nodes that follow irrational choices. But, since the Upfront Money Pump is BI-terminating, the choices that are prescribed by backward induction can be defended by a more compelling argument without this assumption. We only need the weaker assumption that, at nodes that can be reached without making any irrational choices, you retain (i) your rationality and (ii) your trust in your rationality at nodes that can be reached without making any irrational choices. The argument takes the form of a proof by contradiction.[38]

Assume that all three choice nodes can be reached without making any irrational choices. So, at these nodes, you retain your rationality and your trust in your rationality at these nodes. Accordingly, you would accept the trade at node 3, since you prefer C to A. Taking this into account at node 2, the choice at node 2 is effectively between B and C. And, since you prefer B to C, it's irrational to turn down the trade at node 2. This contradicts our assumption that all three choice nodes can be reached without making any irrational choices.

Next, assume that nodes 1 and 2 can be reached without making any irrational choices. So, at these nodes, you retain your rationality and your trust in your rationality at these nodes. Since we have already shown that it's irrational to go down at (at least) one of nodes 1 and 2, it must be irrational to go down at node 2. So it must be rationally required to accept the trade at node 2.[39]

---

[38] Gustafsson and Rabinowicz 2020, p. 585. The argument is adapted from Broome and Rabinowicz 1999, pp. 240–2. See also Rabinowicz 1998, pp. 108–9 and Aumann 1998, p. 103.

[39] This assumes the principle that its being prohibited (forbidden) to ϕ entails its being required (obligatory) not to ϕ. See von Wright 1951a, p. 3; 1951b, p. 37 and Gustafsson 2020, p. 121.

Accordingly, you would accept the trade at node 2. Taking this into account at node 1, the choice at node 1 is effectively between A^- and B. And, since you prefer A^- to B, it's irrational to turn down the trade at node 1. This contradicts our assumption that nodes 1 and 2 can be reached without making any irrational choices.

Hence it's irrational to turn down the trade at node 1. So it's rationally required to accept the first trade from A to A^-. And so you end up with A^- when you could have kept A for free.

For this argument, we only assumed that, at nodes that can be reached without making any irrational choices, you retain (i) your rationality and (ii) your trust in your rationality at nodes that can be reached without making any irrational choices.

If your preferences are robust under a uniform monetary sweetening (for instance, a penny), we can extend the Upfront Money Pump so that you pay an arbitrarily high amount. Consider the decision problem in Figure 5, the *Ruinous Upfront Money Pump*.[40]

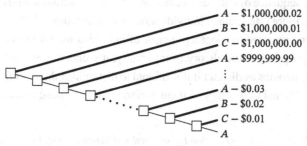

$A \succ B \succ C \succ A$, robust for differences of \$0.01.

Figure 5 The Ruinous Upfront Money Pump

With the same backward-induction argument as before, we find that it's rationally required to accept the first trade. So you end up paying over \$1,000,000 for A when you could have kept A for free. And, as the Ruinous Upfront Money Pump is still a BI-terminating decision problem, we can defend the prescriptions of backward induction in this case with the same minimal assumptions we relied on for the Upfront Money Pump.

It may be objected that choosing A^- at node 1 in the Upfront Money Pump is not a sign of irrationality, since the sequence of choices leading to A isn't available in the relevant sense at that node. The idea being that the sequence of

---

[40] Gustafsson and Rabinowicz 2020, p. 586. Compare Burros (1974, p. 190) who complains that money pumps need only lead to negligible losses for cyclic preferrers. But, as Schick (1986, p. 116) points out, any payment for something the agent could have kept for free is sufficient for the argument. It's the exploitation, rather than the amount lost, that signals irrationality.

choices leading to A isn't securable at node 1, because, at that node, you can't make your future self make those choices.[41]

But the target of the money-pump argument isn't your choice at the first node, which does seem rational *given your preferences*. The target is the structure of your preferences. And the reason why you can't secure the sequence of choices that leads to A (even though you can secure the choice of A^-) is the cyclic structure of your preferences.[42]

It may next be objected that, even though you prefer A^- to B, you could still prefer B to a more specific version of A^-, such as A^--*when-you-could-have-kept-A*.[43] And, if so, you could be rationally required to turn down the trade at node 1 in the Upfront Money Pump even though you prefer A^- to B. Hence, by individuating final outcomes (and thereby prospects) finely enough, you may avoid exploitation even though you have the preferences in (1).

This objection assumes that there are no restrictions on how we may individuate final outcomes. But, if there were no such restrictions, requirements of rationality such as Three-Step Acyclicity would be compatible with the rationality of any sequence of choices, because any alleged violation would disappear given some more fine-grained individuation of final outcomes.[44] To get around this problem, we need to adopt a principle of individuation for final outcomes.

For the purposes of our theory of rationality, it seems that we only need to treat final outcomes as distinct if it's rational to distinguish them preferentially; but, if it is rational to distinguish them preferentially, we need to treat them as distinct:[45]

> *The Principle of Individuation by Rational Indifference*   Final outcomes x and y should be treated as the same if and only if it is rationally required to be indifferent between the sure prospects of x and y.

Of course, if it were rationally permitted not to be indifferent between A and A^--*when-you-could-have-kept-A*, these prospects may still be treated as distinct.

---

[41] Sobel 1976, p. 196, Seidenfeld 1988, p. 278, and Buchak 2013, pp. 180–1.

[42] Steele 2010, p. 474.

[43] Schick 1986, p. 118. Anand (1993a, pp. 62–4; 1993b, pp. 342–3) makes much the same point framed in terms of a counterfactual analysis of preference.

[44] Davidson et al. 1955, p. 145, Tversky 1975, pp. 171–2, and Anand 1990, pp. 94–6.

[45] Broome (1990, p. 140; 1991, p. 103) proposes the following alternative principle:

> *The Principle of Individuation by Justifiers*   Final outcomes x and y should be distinguished if and only if it is rationally permitted to have a preference between the sure prospects of x and y.

> This principle is implausible if it's rationally required to have a preferential gap between some sure prospects. If some final outcomes are so qualitatively different that a preferential gap is required between their sure prospects, they should be treated as distinct. The Principle of Individuation by Rational Indifference avoids this problem.

But, as we shall see in Section 7, it is irrational not to be indifferent between such prospects.[46]

Does the Principle of Individuation by Rational Indifference violate the transitivity of identity (the principle that, if $X = Y = Z$, then $X = Z$)? Consider again the example of having a cup of coffee with one, two, or three lumps of sugar: You can't tell the difference between a cup with one lump and a cup with two lumps. Nor can you tell the difference between a cup with two lumps and a cup with three lumps. But you can tell the difference between a cup with one lump and a cup with three lumps. As before, let A, B, and C be the sure prospects of having a cup with one, two, and three lumps of sugar respectively. When you can't tell the difference between two prospects, it's arguably rationally required to be indifferent between them. So then it's rationally required to be indifferent between A and B and between B and C. Yet it seems rational to have a preference between A and C. So we find that $A = B = C \neq A$, which violates the transitivity of identity. This objection, however, is blocked if we allow indirect ways of telling the difference between the options. You can tell that a cup with one lump (A) tastes noticeably different from a cup with three lumps (C), whereas a cup with two lumps (B) does not taste noticeably different from a cup with three lumps (C). This difference in how they compare to C is a noticeable difference between A and B. And, in the same manner (changing what needs to be changed), you can distinguish B and C.[47]

It may also be objected that you could resist exploitation in the Upfront Money Pump if you adopt self-regulation. Basically, *self-regulation* forbids, if it can be avoided, choosing options that may be followed by a rationally permitted sequence of choices that has a prospect that you would not have chosen in a direct choice between the prospects of all available plans.[48] Following the Uncovered-Choice Rule, you wouldn't choose A^- in a direct choice between A, A^-, B, and C. So self-regulation prescribes that you turn down the first trade in the Upfront Money Pump. And then you avoid exploitation.

Nevertheless, cyclic preferrers who adopt self-regulation are still vulnerable to the arguments we used to defend the choices that backward induction prescribes in the Upfront Money Pump. This is the main problem with the self-regulation defence of cyclicity.[49] Moreover, this objection to self-regulation

---

[46] See the discussion of the Fine-Grained Approach to resolute choice in Section 7.

[47] See Ng 1975, p. 549 and Regan 2000, pp. 51–2 for similar indirect comparisons.

[48] Ahmed 2017, p. 1001. Cubitt and Sugden (2001, p. 143) suggest a similar idea.

[49] Ahmed has two responses to this objection. His (2017, p. 1005) first response is that, even if backward induction shows that self-regulating cyclic preferrers are irrational, such preferrers still avoid exploitation. But this response is irrelevant to our present concern – namely, whether Three-Step Acyclicity is a requirement of rationality. Ahmed's (2017, pp. 1006–9) second response relies on the suggestion that, holding fixed what you would choose at later

works in all the decision problems we will rely on in our overall money-pump argument for Expected Utility Theory.

For a (potential) second way of exploiting cyclic preferrers who rely on self-regulation, we sour all three options in the cycle. From (1), we have, by Unidimensional Continuity of Preference,

(4) $A > A^- > B > B^- > C > C^- > A,$

where A^-, B^-, and C^- are sourings of A, B, and C respectively.

Consider the decision problem in Figure 6, the *Three-Way Money Pump*.[50]

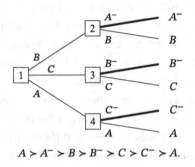

$$A > A^- > B > B^- > C > C^- > A.$$

Figure 6 The Three-Way Money Pump

In the Three-Way Money Pump, you would go up at each of nodes 2, 3, and 4. So, no matter what you choose at node 1, you end up paying for something that you could have had for free. For instance, if you go up at node 1 and up at node 2, then you end up with A^- when you could have had A (by going down at node 1 and down at node 4).

This exploitation scheme doesn't work, however. The exploiter cannot use it without potentially giving up something. The problem is that you can't be relied on to choose a certain option at node 1, so you might end up paying for any one of A, B, and C no matter what your initial holding were. For instance, if you start with A, the exploiter potentially needs to trade you one of B and C to get your money. And then the scheme looks less profitable for the exploiter.[51] It also looks less irrational for you, because you may end up with a final holding that you prefer to your initial holding.[52] If you start off with A and end up with B^- or with C^-, you do not pay for what you could have *kept* for free.

---

choice nodes, you should choose an option with a prospect that would be chosen in a choice between the prospects of all available sequences of choices. This suggestion, however, assumes the point at issue, since it's basically equivalent to self-regulation. Moreover, it is compatible with backward induction given preferences that satisfy Expected Utility Theory.

[50] Cantwell 2002, pp. 328–9; 2003, pp. 388–9.
[51] Ahmed 2017, p. 1011.
[52] Gustafsson and Rabinowicz 2020, p. 589.

It may be objected that whether some behaviour is a sign of irrationality shouldn't depend on how profitable it is for an exploiter. Isn't the mere fact that you chose X^- when you could have had X a sufficient sign of irrationality? If so, our task constructing money pumps would be much easier. But, if that were a sign of irrationality, then the money-pump argument would prove too much, since clearly rational preferences would be irrational in some cases with an infinite series of trades (as we shall see in Section 8). Being exploited by giving exploiters a free lunch seems worrying in a separate way from merely making a sequence of choices that has a prospect that is less preferred than the prospect of some alternative sequence of choices.

Nevertheless, we can modify the set-up so that cyclic preferrers who rely on self-regulation still end up paying for what they could have kept for free. For this variation, we need to sour each option in the cycle once more. From (4), we have, by Unidimensional Continuity of Preference,

(5) $A > A^- > A^{--} > B > B^- > B^{--} > C > C^- > C^{--} > A,$

where A^{--}, B^{--}, and C^{--} are sourings of A^-, B^-, and C^- respectively.

Let a *state of nature* be a description of the world that resolves all of the agent's uncertainty except that it leaves open what the agent will choose.[53] Let an *event* be a set of states of nature.[54] Let $\neg E$ be the complement of event E, that is, the event that E does not occur. Let $E \,\&\, E^*$ be the intersection of events E and E^*, that is, the event that both E and E^* occur. Let a *partition* of states of nature be a set of events such that (i) each event in the set includes at least one state of nature and (ii) each state of nature is a member of exactly one event in the set. And let a *gamble* be a distribution of prospects over a partition of states of nature.

Suppose then that E_1 and E_2 are two independent chance events such that E_1 occurs with a $1/3$ probability and E_2 occurs with a $1/2$ probability. And consider the gambles G_1, G_1^-, and G_2 whose outcomes depend on E_1 and E_2:

	E_1	$\neg E_1 \,\&\, E_2$	$\neg E_1 \,\&\, \neg E_2$
	(1/3)	(1/3)	(1/3)
G_1	A	B	C
G_1^-	A^-	B^-	C^-
G_2	C^{--}	A^{--}	B^{--}

Here, $\{E_1, \neg E_1 \,\&\, E_2, \neg E_1 \,\&\, \neg E_2\}$ is a partition of states of nature.

---

[53] Savage 1954, p. 9, Arrow 1965, p. 12, and Joyce 1999, p. 57.
[54] Savage 1954, p. 10.

We adopt the following requirement of rationality:[55]

> *The Weak Principle of Equiprobable Unidimensional Dominance*   If there
> are sets of events $\{E_1, E_2, \ldots\}$ and $\{E_1^*, E_2^*, \ldots\}$ such that these sets are
> partitions of states of nature and, for all $i = 1, 2, \ldots$, it holds that (a) E_i
> has the same probability as E_i^*, (b) the outcome of gamble G^* given E_i^* is
> a souring of the outcome of gamble G given E_i, and (c) the outcome of G
> given E_i is preferred to the outcome of G^* given E_i^*, then $G > G^*$.

From (5), we have, by the Weak Principle of Equiprobable Unidimensional
Dominance,

(6) $G_1 > G_1^- > G_2$, and $G_1 > G_2$.

Now, consider the decision problem in Figure 7, the *Self-Regulation Money Pump*.

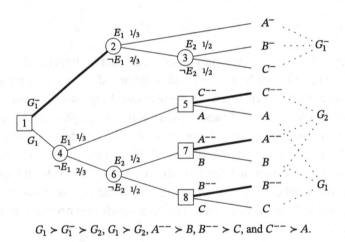

$$G_1 > G_1^- > G_2, G_1 > G_2, A^{--} > B, B^{--} > C, \text{ and } C^{--} > A.$$

Figure 7 The Self-Regulation Money Pump

Here, the circles represent chance nodes where the way forward depends on a
chance event. Chance nodes 2 and 4 go up if and only if E_1 occurs. And chance
nodes 3 and 6 go up if and only if E_2 occurs.

---

[55] The preferences in (5) are also ruled out by the Weak Principle of Equiprobable Unidimensional
Dominance in combination with the following principle:

> *The Weak Principle of Eventwise Dominance*   If there is a set of events such that
> (i) the set is a partition of states of nature and (ii), given each event E in the set, the
> outcome of gamble G given E is preferred to the outcome of gamble G^* given E, then
> $G > G^*$.

From the Weak Principle of Equiprobable Unidimensional Dominance, we obtain that $G_1 >
G_2$. But, from the Weak Principle of Eventwise Dominance, we obtain that $G_2 > G_1$. See
Fishburn 1991, p. 116.

You start off with gamble G_1. At choice node 1, you are offered a trade from G_1 to G_1^-. If you turn down the trade at node 1, you would get an offer to trade from the outcome of G_1 to the outcome of G_2 after the chance events have resolved. This second trade offer would be offered to you at one of choice nodes 5, 7, and 8.

At each of nodes 5, 7, and 8, you would accept the second trade offer. Using backward induction or self-regulation, you take these predictions into account at node 1. And then the choice at node 1 is effectively between G_1^- (accepting the trade) and G_2 (turning it down).

So, taking future choices into account at node 1, neither of the prospects that are effectively available at that node – that is, G_1^- and G_2 – would be chosen in a direct choice between the prospects of all available plans given the Uncovered-Choice Rule, because G_1 is preferred to both of them. Hence self-regulation forbids neither G_1^- nor G_2, since you can't avoid choosing an option with a prospect that you would not have chosen in a direct choice between the prospects of all available plans. So it seems that, even if you rely on self-regulation, you should accept the first trade (since you prefer G_1^- to G_2). But then you end up with G_1^- when you could have kept G_1 for free. Hence this money pump isn't blocked by self-regulation, and it makes you pay for what you could have kept for free.[56]

2.2 Acyclicity

So far, we have only considered arguments that rational preferences conform to Three-Step Acyclicity. We haven't considered the following, more general, requirement:[57]

Acyclicity If $X_1 > X_2 > \ldots > X_n$, then it is not the case that $X_n > X_1$.

Yet we can show that Acyclicity is a requirement of rationality in much the same way as Three-Step Acyclicity.

Suppose that you violate Acyclicity by having the following, arbitrarily large, cycle of preferences:

(7) $A_1 > A_2 > \ldots > A_n > A_1$.

From (7), we have, by Unidimensional Continuity of Preference,

(8) $A_1 > A_1^- > A_2 > \ldots > A_n > A_1$,

---

[56] Moreover, like the Upfront Money Pump, the Self-Regulation Money Pump is BI-terminating.
[57] von Neumann and Morgenstern 1944, pp. 590–1 and Houthakker 1950, p. 162.

where A_1^- is a souring of A_1. We can then extend the Upfront Money Pump to handle this arbitrarily large cycle. Consider the decision problem in Figure 8, the *Upfront Acyclicity Money Pump*.[58]

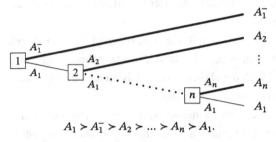

$$A_1 > A_1^- > A_2 > ... > A_n > A_1.$$

Figure 8 The Upfront Acyclicity Money Pump

Here, you are first offered the opportunity to trade from A_1 to A_1^-. If you were to turn down that offer, you would be offered a trade from A_1 to A_2. Then, if you were to turn down that offer too, you would be offered a trade from A_1 to A_3, and so on until you would be offered a final trade from A_1 to A_n.

By backward induction, we find (in the same manner as in the Upfront Money Pump) that it's rationally required to accept the first trade. So you end up with A_1^- when you could have kept A_1 for free.

To spell out the assumptions of the money-pump arguments, we will rely on the notion of plans being available. Let a *plan* at a node n be a specification of what to choose at each choice node that can be reached from n while following the specification. Let us say that one *follows a plan* from a node n if and only if, for each choice node n^* that can be reached from n while choosing in accordance with the plan, one would choose in accordance with that plan if one were to face n^*. Moreover, let us say that one *intentionally follows a plan* from a node n if and only if one follows the plan from n and, for all nodes n^* such that n^* can be reached from n by following the plan, if one were to face n^*, one would either form or have formed at n^* an intention to choose in accordance with the plan at every choice node that can both be reached from n and be reached from n^* by following the plan. Finally, let us say that *a plan is available* at a node n if and only if the plan can be intentionally followed from n.[59]

We assume the following requirement of rationality:[60]

---

[58] Gustafsson and Rabinowicz 2020, p. 584.

[59] Gustafsson 2021, p. 28.

[60] Note that the Principle of Unexploitability is weaker than Hammond's (1988a, p. 508) assumption of *Normal-Form Consequentialism*, which states that one follows a plan from a node if and only if the prospect of that plan would be chosen in a direct choice between the prospects of all available plans at the node. Thoma (2020, pp. 1227–31) argues that the money-pump argument breaks down if it relies on both backward induction and Normal-Form Consequentialism.

The Principle of Unexploitability If (i) X^- is a souring of X, (ii) $X > X^-$, (iii), at node n, it holds that P and P^- are two available plans such that P is the only available plan that amounts to walking away from all offers by an exploiter and the prospect of following P is X and the prospect of following P^- is X^-, and (iv) one knows what decision problem one faces at n, then one does not follow P^- from n.

This is the main assumption of money-pump arguments: that it is irrational to knowingly pay (in some currency you care about) for what you could have kept for free.

We also assume the following principle:

The Principle of Preferential Invulnerability If there is a possible situation where having a certain combination of preferences forces one to violate a requirement of rationality, then there is a requirement of rationality that rules out that combination of preferences in all possible situations.

Given this principle, rational preferences must not lead to any conflicts with any requirements of rationality in any possible situation. The underlying idea is that there's no *rational luck*.[61] Whether you are rational shouldn't depend on what situation you happen to find yourself in. So whether it's unlikely that you will ever face a money-pump set-up is irrelevant.[62]

Putting this together, we have a money-pump argument that Acyclicity is a requirement of rationality, and this argument relies on the following requirements of rationality:

Backward induction, she argues, is plausible given that an instrumental standard of rationality is the *Principle of Open Prospects*, that is, at each time, your end is to bring about the prospect you most prefer then. Whereas, Normal-Form Consequentialism is plausible given that an instrumental standard of rationality is the *Principle of Initial Prospects*, that is, your preferences over the prospects initially open to you define your ends – and the instrumental requirements of rationality require you to do well by your preferences over those prospects. Neither the Principle of Open Prospects nor the Principle of Initial Prospects, however, is sufficient to support both backward induction and Normal-Form Consequentialism. But why can't we adopt both of these instrumental standards of rationality? Thoma's (2020, pp. 1230–1) response is that, if we do, these standards will come into conflict if you violate the strong strict-preference version of Independence with preferences of the kind in (42) at node 4 of the Independence Money Pump (see Section 5.2). Given the Principle of Open Prospects, you should go up at node 4, but, given the Principle of Initial Prospects, you should go down at that node. Yet this response is unconvincing. If you're forced to violate one of these instrumental standards of rationality at node 4, then you are irrational – which is what the money-pump argument aims to show. So why would this show that the argument fails rather than that it succeeds? These two standards are not themselves in conflict; they never conflict for agents whose preferences conform to Expected Utility Theory. So it seems unproblematic to rely on both.

[61] Compare Williams's (1976, p. 116) and Nagel's (1976, pp. 140–6) discussion of the analogous idea of moral luck. Christensen (1996, pp. 457–8) makes a similar point about Dutch-book arguments (see note 9).

[62] Thus we dodge Pettigrew's (2020, p. 58) objection that you (the agent) may find it unlikely that you will ever face this kind of exploitation set-up.

- Backward induction at nodes that can be reached without making irrational choices
- The Principle of Unexploitability
- Unidimensional Continuity of Preference

And, in addition, the argument relies on the following principles:

- Decision-Tree Separability
- The possibility of the Upfront Acyclicity Money Pump
- The Principle of Preferential Invulnerability

We need the possibility of the Upfront Acyclicity Money Pump, since the Principle of Preferential Invulnerability only covers possible situations. This assumption is substantial, since it may be rejected if some outcomes cannot occur in the relevant sequential patterns. We will also take it as part of the description of the Upfront Money Pump (and the other decision problems we will discuss) that, at each node, all plans at that node are available. We do not, however, assume that the money-pump situations we discuss are likely to arise. As conceived here, the money-pump argument against cyclic preferences does not aim to show that having acyclic preferences is useful, or that cyclic preferences are likely to have bad effects.[63]

One worry about the Upfront Acyclicity Money Pump (and the other money pumps we'll discuss) is that that decision problem is impossible if the violating preferences are non-practical preferences. Consider the following example.[64] Let A be the sure prospect of staying at home. Let B be the sure prospect of going to Rome. And let C be the sure prospect of going mountaineering. Here, your preferences may plausibly be sensitive to what alternatives are available. Suppose that you prefer A-when-the-only-alternative-is-B to B, prefer B to C, and prefer C to A-when-the-only-alternative-is-C. These transitive (and plausible) preferences are *practical* in the sense that, for each pairwise preference, there is a possible choice between the compared prospects. Accordingly, preferences are *non-practical* if and only if it is not the case that, for each pairwise

---

[63] Therefore, Parfit's (2011, p. 128) objection that cyclic preferences may also have good effects does not apply to money-pump arguments as devised here. Nor does Halstead's (2015, p. 204) objection that bad practical results needn't be a sign of irrationality. Having said that, we may still wonder whether cyclic preferrers would suffer bad effects from money pumps. Etchart (2002, p. 22) argues that, since most people are dynamically inconsistent, they wouldn't subject others to what they wouldn't like to undergo from them. But they could still be money pumped by dynamically consistent exploiters and exploiters who don't live by the golden rule. Etchart (2002, p. 22) also suggests that, given free entry on a competitive market, competition among exploiters would bring down the price. Nevertheless, while the competition would bring down the individual profit for the exploiters, the exploitees would still lose their money.

[64] Broome 1993, pp. 53–4.

preference, there is a possible choice between the compared prospects. But suppose that you also prefer *C* to *A-when-the-only-alternative-is-B*. This makes your preferences cyclical. So, to defend Acyclicity, we need to show that these preferences are irrational. The additional preference, however, is non-practical, since there is no possible choice between *C* and *A-when-the-only-alternative-is-B*. So, for these cyclic preferences, the Upfront Acyclicity Money Pump is impossible.

To handle this problem, we rely on the Principle of Individuation by Rational Indifference.[65] The fine-grained prospects *A-when-the-only-alternative-is-B* and *A-when-the-only-alternative-is-C* differ not only in what the alternative would be but also in that you are a coward in the latter but not in the former. Plausibly, the difference you are rationally permitted to care about is not the difference in alternative but the difference in cowardice. So you are rationally required to be indifferent between *A-when-the-only-alternative-is-B* and *A-and-not-being-a-coward*. The Principle of Individuation by Rational Indifference then entails that these prospects should be treated as the same. Crucially, it's possible to have a choice between *A-and-not-being-a-coward* and *C*. So there is a possible instance of the Upfront Acyclicity Money Pump for your following practical preference cycle: *A-and-not-being-a-coward* is preferred to *B*, *B* is preferred to *C*, and *C* is preferred to *A-and-not-being-a-coward*.[66]

What about cycles with fewer steps than three? Consider the following irreflexivity requirement:[67]

---

[65] See also the discussion of the Fine-Grained Approach to resolute choice in Section 7.

[66] For another worry about the possibility assumption, consider the cycle of preferences in the Mere-Addition Paradox, where *A* is a sure prospect with a large number of lives with very high positive quality, *A+* is a sure prospect that is just like *A* except that (in addition to the individuals in *A*) there are very many additional lives with minimally positive well-being, and *Z* is a sure prospect where the same individuals exist as in *A+* and their lives are barely worth living but are slightly better than the worst lives in *A+* so that there is more well-being overall. Many people have the preferences $A > Z > A+ > A$. (See McMahan 1981, pp. 122–3 and Parfit 1982, pp. 158–60.) Yet consider a version of origin essentialism, where each person has this necessary property: that of having been brought into existence by the particular causal history that in fact brought them into existence. (This version of origin essentialism differs from the one defended in Kripke 1972, pp. 350–351n56.) Given this causal-history version of origin essentialism, there couldn't be an instance of the Upfront Acyclicity Money Pump for these cyclic preferences, because (however the outcomes are arranged) there would be multiple alternative causal histories that bring about the existence of the additional people who exist in both *A+* and *Z* but not in *A*. But origin essentialism is dubious. It seems that what is essential to individuals should be symmetrical with respect to the direction of time. If the fundamental laws of nature are invariant with respect to the direction of time (see Feynman et al. 1963, ch. 52, p. 3), it seems that the fundamental metaphysical laws should be so too. Consider *terminus essentialism*, the claim that there is only one possible way you might die, namely, the way you will in fact die. Terminus essentialism is implausible. And, since we reject terminus essentialism, we should reject origin essentialism too.

[67] Uzawa 1956, p. 35 and Hansson 2002, p. 325.

One-Step Acyclicity   It is not the case that $X > X$.

And consider the following asymmetry requirement:[68]

Two-Step Acyclicity   If $X > Y$, then it is not the case that $Y > X$.

Whether it's even possible to violate these principles depends on how we define strict preference. Let '$X \succsim Y$' denote that X is at least as preferred as Y.[69] We then adopt the following definition of that X is *preferred* to Y:[70]

$X > Y =_{df} X \succsim Y$ and it is not the case that $Y \succsim X$.

Given this definition of strict preference, violations of One- and Two-Step Acyclicity are impossible.[71]

3 Completeness

Consider having an apple or having an orange. Suppose that, given their different qualities, you can't compare these options: you do not prefer one of them to the other, yet you're not indifferent between them.[72]

Your preferences violate Completeness, the first basic axiom of Expected Utility Theory:[73]

Completeness $X \succsim Y$ or $Y \succsim X$.

We distinguish between indifference, which does not violate Completeness, and a preferential gap, which does. Let '$X \sim Y$' denote that X is *indifferent* to Y, defined as follows:[74]

---

[68] Halldén 1957, p. 25, Uzawa 1960, p. 134, and Hansson 2002, p. 325.

[69] Herstein and Milnor 1953, p. 292.

[70] Debreu 1954, p. 160.

[71] If we dropped this definition, we could defend these requirements of rationality by contracting the Upfront Acyclicity Money Pump with $n = 1$ and with $n = 2$. Still, it's hard to grasp what violations of these requirements would be like.

[72] There are many variations of this example. In Sartre's (1946, pp. 39–47; 2007, pp. 30–3) variation, the two options are caring for your mother and fighting for your country. Aumann (1962, p. 446) offers a different kind of example, where you prefer a cup of cocoa to a lottery between a cup of coffee and a cup of tea if the probability of coffee is high, but you prefer a lottery between coffee and tea to cocoa if the probability of coffee is low. Yet there needn't be an exact probability of coffee at which your preference changes direction and at which you're indifferent between cocoa and the lottery between coffee and tea. There would be a range of probabilities of coffee at which you have a preferential gap between the lottery and the cocoa.

[73] Arrow 1951, p. 13. Completeness should be distinguished from the following weaker principle:

Connectedness   If $X \neq Y$, then $X \succsim Y$ or $Y \succsim X$.

See von Neumann and Morgenstern 1944, pp. 26–7, Arrow 1951, p. 13n6, and Roberts 1979, p. 15. (In the literature, however, 'Completeness' and 'Connectedness' are used interchangeably.)

[74] Arrow 1951, p. 14. The notation comes from Herstein and Milnor 1953, p. 292.

$X \sim Y =_{df} X \gtrsim Y$ and $Y \gtrsim X$.

Let '$X \parallel Y$' denote a *preferential gap* between X and Y, defined as follows:[75]

$X \parallel Y =_{df}$ it is neither the case that $X \gtrsim Y$ nor the case that $Y \gtrsim X$.

Let A be the sure prospect of having the apple, and let B be the sure prospect of having the orange. Then we can state your preferences as follows:

(9) $A \parallel B$.

As defined, indifference and preferential gaps are both symmetrical relations.[76] So, in the absence of any further requirements of rationality, it's hard to make any practical distinction between these relations. If we, for example, take some preferences that satisfy Expected Utility Theory and replace all indifference relations with preferential gaps, the resulting preferences would be no more exploitable than the original preferences. These new preferences, which violate Completeness, would be practically equivalent to the original preferences, which satisfy Completeness. So we need some practically relevant difference between indifference and preferential gaps or the distinction won't matter – robbing Completeness of practical substance.

A plausible distinguishing feature of preferential gaps is their insensitivity to at least some sourings (and to at least some sweetenings):[77]

> *Weak Insensitivity to Souring*   If $X \parallel Y$, then
> - there is a prospect X^- such that (i) X^- is a souring of X and (ii) $X > X^- \parallel Y$ or
> - there is a prospect Y^- such that (i) Y^- is a souring of Y and (ii) $Y > Y^- \parallel X$.

The idea is that this robustness to sourings holds for preferential gaps but not for indifference. But couldn't two prospects be related by a preferential gap even though any souring of either prospect breaks the gap? For the purpose of our discussion, we can treat such prospects as being indifferent, because the main assumption we will make about indifference for the money-pump argument for Transitivity (specifically, the souring approach in Section 4.1) is that any

---

[75] von Neumann and Morgenstern 1947, p. 630.

[76] A relation over a set is *symmetrical* if and only if, for all x and y in the set, if x is related to y, then y is related to x. See De Morgan 1851, p. 104 and Russell 1903, p. 218.

[77] Compare Raz's (1985–6, p. 120; 1986, pp. 325–6) similar 'mark of incommensurability' for value incomparability. Broome (2004, p. 21) suggests another way to tell apart equal goodness from value incomparability – namely, that 'X is equally good as Y' means that X is neither better nor worse than Y and, for any Z, it holds that Z is better or worse than X if and only if Z is correspondingly better or worse than Y. Broome adopts this definition to ensure the transitivity of 'equally good as' (see note 103). So we wouldn't want to rely on a preferential analogue of his definition, since we aim to defend Transitivity (and not to make it true by definition).

souring would break the indifference between prospects. Hence we treat Weak Insensitivity to Souring as a stipulation rather than an assumption.

We will, however, make the substantial assumption that the following principle is a requirement of rationality:

> *Symmetry of Souring Sensitivity*   If (i) X^- is a souring of X and (ii) $X >$ $X^- \parallel Y \parallel X$, then there is a prospect Y^- such that (i) Y^- is a souring of Y and (ii) $Y > Y^- \parallel X$.

If you have a preferential gap between X and Y, there must be some kind of perplexity about the comparison of these prospects. This perplexity should plausibly be symmetrical – in the sense that, if the perplexity swallows sourings on one side, it should also do so on the other.[78]

Next, given Weak Insensitivity to Souring and that Symmetry of Souring Sensitivity is a requirement of rationality, we derive the following requirement of rationality:

> *Strong Insensitivity to Souring*   If $X \parallel Y$, then there is a prospect X^- such that (i) X^- is a souring of X and (ii) $X > X^- \parallel Y$.

From (9), we have, by Strong Insensitivity to Souring,

(10) $A > A^- \parallel B \parallel A$,

where A^- is a souring of A.

Now, consider the (potential) money pump for preferential gaps in Figure 9, the *Single-Souring Money Pump*.[79]

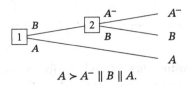

$$A > A^- \parallel B \parallel A.$$

Figure 9 The Single-Souring Money Pump

At node 2, you have a preferential gap between the two options, A^- and B. It seems, therefore, that it's neither irrational to choose A^- nor irrational to choose B.[80]

---

[78] It may be objected that some prospects need not have any sourings. Suppose that you can't compare existence with non-existence. Accordingly, you have a preferential gap between a prospect where you exist and a prospect where you never exist. (See Williams 1973, p. 87.) It seems that there needn't be any sourings of the latter prospect. Nevertheless, examples involving non-existence are irrelevant if we restrict Expected Utility Theory (and Symmetry of Souring Sensitivity) to prospects whose potential final outcomes could be outcomes in at least some decision problems you might face.

[79] Chang 1997, p. 11.

[80] Why not say that it's rationally permitted to choose either option? I am distinguishing between something's being rationally permitted and something's not being irrational (that is, not being

So, at node 1, backward induction does not let you rule out any of the options being chosen (or picked) at node 2.[81] But, if you can't rule out that any one of the options at node 2 would be chosen, it's unclear how you should take that choice into account at node 1.[82] One of the options at node 2, *B*, is no less preferred than the prospect of going down at node 1, *A*. So it may seem that it shouldn't be irrational to go up at node 1. And, if it isn't irrational to go up at node 1, it seems that it isn't irrational to both go up at node 1 and go up at node 2. But, if you go up at both node 1 and node 2, you end up with A^- when you could have kept *A* for free, which violates the Principle of Unexploitability.

The Single-Souring Money Pump, on this interpretation, is an example of a non-forcing money pump. A money-pump set-up is *forcing* if and only if the agent is rationally required, at each step of the set-up, to going along with the exploitation. A money-pump set-up is *permitting* if and only if, at each step of the set-up, the agent is rationally permitted to go along with the exploitation. A money-pump set-up is *non-prohibiting* if and only if, at each step of the set-up, the agent is not rationally prohibited from going along with the exploitation. Finally, a money-pump set-up is *non-forcing* if and only if it is non-prohibiting and, at some step of the set-up, the agent is not rationally required to go along with the exploitation.[83]

---

rationally prohibited). Consider a choice between *X* and *Y*. If *X* is preferred to *Y*, it seems that it's rationally permitted to choose *X* but not rationally permitted to choose *Y*. If *X* is indifferent to *Y*, it seems that it's rationally permitted to choose *X* and rationally permitted to choose *Y*. You might think that the same holds if there is a preferential gap between *X* and *Y* – that it's rationally permitted to choose either prospect. (See, for instance, Peterson 2007, p. 507.) I think, however, that, if there is a preferential gap between *X* and *Y*, then it's not rationally permitted to choose *X* and not rationally permitted to choose *Y* but it's neither irrational to choose *X* nor irrational to choose *Y*. I adopt the following definitions of rationally required and irrational:

> It is *rationally required* to ϕ =df it is rationally permitted to ϕ and it is not the case that it is rationally permitted not to ϕ.
> It is *rationally prohibited (irrational)* to ϕ =df it is rationally permitted not to ϕ and it is not the case that it is rationally permitted to ϕ.

(See Gustafsson 2020, pp. 123–4.) Given these definitions, an option can be (i) rationally permitted, (ii) rationally prohibited, or (iii) neither rationally permitted nor rationally prohibited. Still, in a choice between *X* and *Y*, the difference between *that each of X and Y is rationally permitted* and *that neither of X and Y is rationally permitted* won't matter for our discussion. What matters for our discussion is that, in either case, neither option is irrational.

[81] Ullmann-Margalit and Morgenbesser (1977, p. 757) make a distinction between picking and choosing: If you are indifferent or have a preferential gap between two options, then you can merely pick, not choose, one of them. For our discussion, we collapse this distinction so that picking also counts as choosing.

[82] Peterson (2007, pp. 507–9) argues that it must be rationally permitted to accept the first trade, but his argument implicitly assumes that the agent reasons in a myopic manner. It can be blocked if the agent uses foresight.

[83] The distinction between forcing and non-forcing money pumps is due to Gustafsson and Espinoza (2010, pp. 761–2). See also Gustafsson 2010, p. 252.

While non-forcing money pumps may be problematic for the agent, they are implausible as exploitation schemes.[84] Since it's not irrational to choose B at node 2 of the Single-Souring Money Pump, you might turn down the second trade. And, if you do, you end up with B and the exploiter has given up B for A. So, even though the Single-Souring Money Pump does offer the exploiter an opportunity to potentially get your money for free, the exploiter might end up merely trading you B for A – which need not be in their interest (nor is it contrary to your interest).

Yet, if you make the sequence of choices consisting in accepting both trades, you still violate the Principle of Unexploitability, which (we have assumed) is a requirement of rationality. What is the relationship between the rational status of a sequence of choices and the rational status of the individual choices in that sequence? Consider the following principle:[85]

> *The Principle of Rational Decomposition* If an agent, whose credences and preferences are not rationally prohibited, makes a sequence of choices which violates a requirement of rationality, then some of those choices are rationally prohibited.

This principle is plausible. If no choice in a sequence of choices is irrational, it's hard to see where the irrationality of the sequence would be coming from (given that your credences and preferences aren't irrational). If you didn't violate any requirement of rationality at any point during an interval, it seems that you didn't violate any requirement of rationality during the interval.

Suppose that, contrary to the Principle of Rational Decomposition, you make an irrational sequence of choices where no choice is irrational and your credences and preferences are not irrational. Then this sequence is only ruled

---

[84] Another worry is that non-forcing money pumps may be blocked by additional requirements that don't conflict with the agent's preferences. But, given further assumptions about the agent, this loophole can be closed. For instance, in Section 2.1, we used a non-forcing but still permitting money pump against cyclic preferrers who rely on naive choice and the Uncovered-Choice Rule.

[85] Elga (2010, pp. 9–10) and Tadros (2019, pp. 198–202) both defend similar principles. Note that the Principle of Rational Decomposition is not open to the standard counter-examples to the principle that, if ϕ & ψ is prohibited, then ϕ is prohibited or ψ is prohibited. Suppose that, in a single choice situation, you have a choice whether to ϕ and whether to ψ. And suppose that doing both ϕ and ψ has the worst outcome, doing neither ϕ nor ψ has a better outcome, and doing exactly one of ϕ and ψ has the best outcome. Finally, suppose that you will actually do neither ϕ nor ψ. Then it seems that, given a binary form of consequentialism (where doing something is compared with not doing it), it is prohibited to ϕ & ψ even though it is permitted to ϕ and permitted to ψ. (See Goldman 1978, p. 186 and Gustafsson 2018, p. 102.) This kind of example does not work in case you first have a choice whether to ϕ and then a choice whether to ψ. Other alleged counter-examples, such as Quinn's (1990, pp. 79–80) puzzle of the self-torturer, depend on cyclic preferences that can be shown to be irrational with the money-pump argument in Section 2. So those examples aren't counter-examples to the Principle of Rational Decomposition.

out by diachronic requirements of rationality – that is, the sequence is irrational but no requirement of rationality rules out your credences, preferences, or choices at any moment during the sequence. So, during the interval in which you made this sequence of choices and violated these diachronic requirements, there was no moment at which you did something that violated any requirement of rationality. This robs the prohibition of the sequence of any practical relevance, because we cannot make atemporal choices. So, without any help from other requirements, how could these diachronic requirements guide you away, practically, from completing the sequence?[86] It is tempting to say that, at the final choice node where you have a choice whether to complete the irrational sequence of choices, these diachronic requirements would be violated if you were to make that final choice of the sequence; so that final choice must be rationally prohibited.[87] But, if so, we have no violation of the Principle of Rational Decomposition.[88]

Does the Principle of Rational Decomposition conflict with the Principle of Unexploitability? You might violate the latter by following a dominated plan, where each choice seems rational given your preferences. But that violation of the Principle of Unexploitability need not violate the Principle of Rational Decomposition, since your preferences could be irrational. And, if your preferences are irrational, you violate a requirement of rationality at each moment you have those preferences.

Given the Principle of Rational Decomposition and the Principle of Unexploitability, it's either irrational to go up at node 1 or irrational to go up at node 2 (assuming that your credences and preferences aren't irrational). Could we plausibly claim that it's irrational to go up at node 2?

---

[86] Note that I do not question the existence of diachronic requirements of rationality. I just question the existence of diachronic requirements of rationality which could be violated without violating any requirement of rationality at any moment.

[87] Tadros 2019, p. 201.

[88] Rabinowicz (2012, p. 145) proposes an alleged counter-example. Suppose that we model incomplete preferences (that is, preferences with preferential gaps) as the intersection of the set of complete preference orderings between which the agent is undecided. And suppose that there are two preferences orderings in the set: (i) $B > A > A^-$ and (ii) $A > A^- > B$. Then, in the Single-Souring Money Pump, Rabinowicz claims that it is not rationally prohibited to accept the trade from A to B at node 1, since B is preferred to A in (i). And it is not rationally prohibited to accept the trade from B to A^- at node 2, since A^- is preferred to B in (ii). But the sequence of accepting both trades is rationally prohibited, since A is preferred to A^- in both (i) and (ii). Given that the agent's credences and preferences are rationally permitted, this contradicts the Principle of Rational Decomposition. As stated, however, the example relies on myopic choice, which is implausible. Rabinowicz suggests (in a personal communication) that we amend the example by supposing that the agent has a high credence in (the falsehood) that B would be chosen. The trouble with this alleged counter-example is that it is still open to the main argument for the Principle of Rational Decomposition: the mere irrationality of the sequence does not, at any moment, guide the agent away from completing the sequence.

A potential way to do so is to adopt forward induction. With *forward induction*, one deliberates under the assumption that one's past choices were rational.[89] If the choice to go up at node 1 were rational (or at least not irrational), it seems that you must choose B at node 2. You must choose B at node 2, because, if you instead choose A^-, then the choice to go up at node 1 was effectively a choice of A^- when you could have kept A. And to choose A^- when you could keep A is irrational, contradicting the assumption that the choice at node 1 wasn't irrational.[90]

But forward induction based on your own choices is implausible. The trouble lies in explaining why the rational status of your choice at node 1 should matter to you at node 2.[91] The reason why going up at node 1 would be irrational if you were to also go up at node 2 is that you then end up with A^- when you could still have kept A at node 1. At node 2, however, A is no longer feasible. That A wasn't chosen at node 1 is now (at node 2) just a sunk cost.[92] The only thing that should guide your choice at node 2 is your preference between the still feasible options. That is, we rely on Decision-Tree Separability.

Since it's implausible that choosing A^- would be irrational at node 2, let us turn to the other alternative. Can we plausibly claim that it's irrational to go up at node 1? We can.

Suppose that you went up at both node 1 and node 2. Then you end up with A^-, which is less preferred than something you could have chosen at node 1, namely, A. What choice do you regret? It should be the first choice. With the second choice you merely turned down a prospect that you don't prefer to the prospect you ended up with. So you have no reason to regret the

---

[89] Kohlberg and Mertens 1986, p. 1013. On the standard game-theoretic definition of forward induction, one makes deductions on the assumption that past moves of other players were rational (rather than one's own past moves).

[90] In Gustafsson 2016, p. 66n18, I argued similarly that choosing A^- at node 2 is not irrational but this choice makes the earlier choice at node 1 irrational and that the choice at node 1 would have been rationally permitted if you had chosen B at node 2. (See Bader 2019, p. 246 for much the same suggestion.) But this makes the rational status of the options at node 1 depend on future events. It implausibly violates the Principle of Future-Choice Independence (defined later).

[91] If you use the standard game-theoretic form of forward induction (see note 89), you assume that the past moves of *other* players were rational and this assumption matters to you because it helps you predict what they will do in the future. This may in principle be applicable even to your own past moves insofar as those choices give you some information about what you are going to do in the future. But, at node 2, there is no future to worry about, since you're making a terminal choice.

[92] Machina (1989, p. 1653) tries to rebut the charge of the sunk-cost fallacy, but he only shows that the utility of the current options can depend on earlier choices. He does not show that the value of current options depends on what could have been but was not chosen earlier, which is what the charge targets. For a discussion of the related Fine-Grained Approach to resolute choice, see Section 7.

second choice. Looking back, it's with the first choice that you turned down what you prefer to your final holding.[93]

Given that going up at node 1 is irrational if you also go up at node 2, it seems that going up at node 1 should be irrational regardless of what you end up choosing at node 2. Whether it's rational to choose a certain option at a node or whether it's rational to have certain preferences or credences at that node shouldn't depend on what would in fact happen at later nodes; it should only depend on the state of the world at the time of the choice (and, possibly, earlier times). Whether it's rational to choose a certain option at a node may, of course, depend on the agent's *credences* about what would be chosen at later choice nodes. I'm only denying that what it's rational to choose now could depend on what will actually happen in the future. We accept the following principle:

> *The Principle of Future-Choice Independence* The rational status of an option at a choice node and the rational status of the agent's credences and preferences at that node do not depend on what would in fact be chosen at later choice nodes.

Note that this principle does not conflict with backward induction, since backward induction only relies on *predictions* about what would be chosen at later choice nodes and not on what would in fact be chosen.

Since going up at node 1 is irrational if you also go up at node 2, it follows, by the Principle of Future-Choice Independence, that it's irrational to go up at node 1 no matter what you would choose at node 2.

This argument that it's irrational to go up at node 1 can be generalized. We will do so now to show that the following principle is a rational requirement:

> *Minimal Unidimensional Precaution* If (i) X^- is a souring of X, (ii) $X >$ X^-, (iii) it is not the case that $Y > X^-$, (iv) node n is a choice between node n^* and X, (v) node n^* is a choice between X^- and Y, and (vi) one knows at node n what decision problem one faces, then one chooses X at node n.

We noted earlier that the sequence of choices consisting in going up at both choice nodes in the Single-Souring Money Pump is irrational since it violates the Principle of Unexploitability. Now we assume, more generally, the following principle:[94]

---

[93] Loomes and Sugden (1987, p. 286) argue that if you avoid paying for options such that you would regret having turned down getting them for free, then you avoid being money pumped. The problem is that the choice you would regret is not the choice at node 2; it would be the earlier choice at node 1. Since you know that you won't regret the choice at node 2, it seems that you wouldn't be rationally required at that node to avoid regretting that you did not keep *A* for free.

[94] Broome 1999, p. 156; 2000, p. 33. This principle is more general than the earlier claim, since it is stipulated in the Single-Souring Money Pump that *A* is the walk-away option. The Irrationality of Single-Souring entails that the sequence that ends with A^- would be irrational even if the sequence that ends with *B* were the walk-away option.

> *The Irrationality of Single Sourings*   If (i) X^- is a souring of X, (ii) $X > X^-$, (iii) node n is a choice between node n^* and X, (iv) node n^* is a choice between X^- and Y, and (v) one knows at node n what decision problem one faces, then the sequence of choices consisting in choosing node n^* at node n and X^- at node n^* violates a requirement of rationality.

Suppose that you violate Minimal Unidimensional Precaution by, for instance, going up at node 1 of the Single-Souring Money Pump and having the following preferences, which are entailed by the preferences in (10):

(11) $A > A^-$, and it is not the case that $B > A^-$.

Assume, for proof by contradiction, that you did not violate a requirement of rationality even though you violated Minimal Unidimensional Precaution. So your credences and preferences are not rationally prohibited. Now, regardless of whether you will in fact choose A^- at node 2, we may consider what the rational status of your choices would be if you were to choose A^- at node 2. Note first that, even if you actually choose B at node 2, your credences and preferences at nodes 1 and 2 would be the same as they actually are at these nodes if you were to choose A^- at node 2.[95] From the Principle of Future-Choice Independence, it then follows that your credences and preferences wouldn't be rationally prohibited if you were to choose A^- at node 2. So, if you were to go up at both choice nodes, it follows, by the Irrationality of Single Sourings, that this sequence of choices would be irrational. So, by the Principle of Rational Decomposition, at least one of your choices would be irrational (since your credences and preferences are not rationally prohibited). But, given Decision-Tree Separability, your choice at node 2 cannot be irrational. Hence, if you were to go

---

[95] This follows from

> *The Fixity of the Macro Past*   If an agent who can choose otherwise at node n were to choose otherwise at n, then all macrofacts about the world up to the time of n would be the same as they actually are.

Here, following Aaronson (2016, p. 246), a *macrofact* is a fact that could, in principle, be detected by an external measuring device without disturbing the physical system in any significant way. An agent's credences and preferences are macrofacts. Aaronson (2016, p. 246) distinguishes macrofacts from *microfacts* – facts about undecohered quantum states. Aaronson (2016, pp. 219–20) argues that, if we were to act otherwise, then some microfacts (but no macrofacts) would have been different than they actually are. The following principle is more standard than the Fixity of the Macro Past:

> *The Fixity of the Past*   If an agent who can choose otherwise at node n were to choose otherwise at n, then all facts about the world up to the time of n would be the same as they actually are.

(See van Inwagen 1983, p. 96.) Many arguments for this principle – such as Ginet's (1990, pp. 109–10) – only support the Fixity of the Macro Past, because they rely on the plausibility of holding fixed known macrofacts about the past.

up at both choice nodes, it would be your choice to go up at node 1 that would be irrational. Then, by the Principle of Future-Choice Independence, it follows that the rational status of your choice at node 1 cannot depend on what you choose at node 2. Hence the choice to go up at node 1 must be irrational regardless of whether you would choose B at node 2. So the choice to go up at node 1 is irrational. And, since this argument can be given for all violations of Minimal Unidimensional Precaution (changing what needs to be changed), it follows that Minimal Unidimensional Precaution is a requirement of rationality.

Hence – from Decision-Tree Separability, the Irrationality of Single Sourings, the Principle of Future-Choice Independence, and the Principle of Rational Decomposition – we have derived that Minimal Unidimensional Precaution is a requirement of rationality.

This result may seem puzzling in case you're certain what you would choose at a future node even though the choice at that node is a choice between two options that are related by a preferential gap. You could, for instance, be certain that you will follow a particular tie-breaker rule for resolving choices where neither option is preferred to the other.[96] It may be objected that, if you are certain at node 1 that you would follow a tie-breaker rule at node 2 which favours choosing B, then you seem rationally permitted to go up at node 1, even though this violates Minimal Unidimensional Precaution. If you are certain in this manner what you would choose in the future, you do not (according to this objection) need precaution.

But note that the argument for Minimal Unidimensional Precaution makes no assumptions about what your credences are. So the argument's assumptions should be no less plausible in case you're certain at node 1 that B would be chosen at node 2. The source of puzzlement, here, may be the existence of an objection in the vicinity which does block the argument for Minimal Unidimensional Precaution: It may seem that, if you are certain at node 1 that you would follow a particular tie-breaker rule at node 2, then it would be irrational not to follow that tie-breaker rule at node 2. This suggestion contradicts one of the argument's assumptions – namely, Decision-Tree Separability. So this suggestion would block argument. We will take Decision-Tree Separability for granted, however, until Section 7. So we postpone our discussion of this objection until then.[97]

While we will only need Minimal Unidimensional Precaution for the money-pump argument for Completeness, we can defend a more general precautionary

---

[96] Seidenfeld 2000, p. 306 and Rabinowicz 2000b, pp. 314–15.

[97] That is, in Section 7, we will discuss the Conservative Approach to resolute choice as a requirement of rationality.

form of backward induction with a slightly less conclusive argument. Let a *rationally allowed outcome* of an option X be a prospect of an available plan consisting in choosing X followed by choices that are not irrational. At node 1, there's no potential upside to going up, because no rationally allowed outcome of going up is preferred to the prospect of going down. But there is a potential downside to going up, because one of the rationally allowed outcomes of going up is less preferred than the prospect of going down. So it's irrational to go up at node 1. This argument supports a precautionary version of backward induction:

> According to *precautionary backward induction*, it is irrational to choose an option X over an option Y if there is a rationally allowed outcome of X (that is, a prospect of an available plan consisting in choosing X followed by choices that are not irrational) that is less preferred than some rationally allowed outcome of Y and there is no rationally allowed outcome of Y that is less preferred than some rationally allowed outcome of X.

Precautionary backward induction entails (given Decision-Tree Separability) that going up is irrational at node 1. Likewise, if Minimal Unidimensional Precaution is (as we have argued) a requirement of rationality, we also find that going up is irrational at node 1. So the attempted exploitation in the Single-Souring Money Pump is blocked.

Hence following either Minimal Unidimensional Precaution or precautionary backward induction makes you invulnerable to the Single-Souring Money Pump, but, as we shall see, it does not save you from two variations of that decision problem.

From (10), we have, by Strong Insensitivity to Souring,

(12) $A > A^- \parallel B > B^- \parallel A \parallel B$,

where B^- is a souring of B.

Consider the decision problem in Figure 10, the *Dual-Souring Money Pump*.[98]

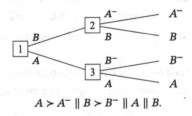

$A > A^- \parallel B > B^- \parallel A \parallel B.$

Figure 10 The Dual-Souring Money Pump

---

98 Hammond 1988b, p. 295.

No matter what you choose at node 1 in this case, you will end up at a node where (given Decision-Tree Separability) it is not irrational to pay for something you could have had for free. And neither Minimal Unidimensional Precaution nor precautionary backward induction will help you.

But the Dual-Souring Money Pump is not a plausible exploitation scheme, because the exploiter might end up merely trading you B for A or A^-. This wouldn't necessarily be in the exploiter's interest, nor would it be contrary to your interest. (Hence the Dual-Souring Money Pump suffers from the same problem as the Three-Way Money Pump.)

Nevertheless, we can construct a forcing money-pump set-up against agents with incomplete preferences, given Minimal Unidimensional Precaution (or precautionary backward induction). From (12), we have, by Strong Insensitivity to Souring,

(13) $A > A^- > A^{--} \parallel B > B^- \parallel A \parallel B \parallel A^-$,

where A^{--} is a souring of A^-.

Now, consider the decision problem in Figure 11, the *Precaution Money Pump*.

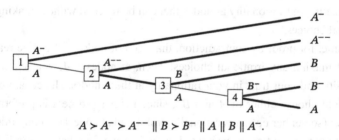

$$A > A^- > A^{--} \parallel B > B^- \parallel A \parallel B \parallel A^-.$$

Figure 11 The Precaution Money Pump

At node 4, neither option is less preferred than the other, so it's neither irrational to choose A nor irrational to choose B^-.

Taking this into account at node 3, we find that turning down the trade at node 3 has a rationally allowed outcome (B^-) that is less preferred than the prospect of accepting the trade (B) but there is no rationally allowed outcome of turning the trade down that is preferred to the prospect of accepting the trade. So, by precautionary backward induction, you would accept the trade at node 3. Alternatively, we could rely on Minimal Unidimensional Precaution, which also prescribes going up at node 3 (since B^- is a souring of B and you do not prefer A to B^-).

Taking this prediction into account at node 2, the choice at that node is effectively between A^{--} and B. Since neither of these options is less preferred than the other, it's neither irrational to accept nor irrational to turn down the trade at node 2.

Taking this into account at node 1, we find that turning the trade down has a rationally allowed outcome (A^{--}) that is less preferred than the prospect of accepting the trade (A^-) but turning the trade down cannot lead (via choices that aren't irrational) to a prospect that is preferred to the prospect of accepting the trade. So, by precautionary backward induction, you accept the trade at node 1. Alternatively, we could rely on Minimal Unidimensional Precaution, which also prescribes going up at node 1 (since A^{--} is a souring of A^- and you do not prefer B to A^{--}). Hence you accept the first trade and end up with A^-, even though you could have kept A for free.

The Precaution Money Pump isn't BI-terminating, since going down at node 2 is allowed by backward induction and would be followed by the choice at node 3. Even so, we can still defend the prescriptions of precautionary backward induction or Minimal Unidimensional Precaution combined with standard backward induction without assuming that you retain your rationality and your trust in your rationality at nodes that can only be reached by irrational choices. As before, we only need the assumption that, at nodes that can be reached without making any irrational choices, you retain (i) your rationality and (ii) your trust in your rationality at nodes that can be reached without making any irrational choices.

Assume, for proof by contradiction, that each of nodes 2–4 can be reached without making any irrational choices. Then, at nodes 1–4, you retain your rationality and your trust in your rationality at these nodes. Hence, at node 4, you might choose either of A and B^-, since neither prospect is preferred to the other (so neither option is rationally prohibited at node 4). Taking this into account with precautionary backward induction, we find that it's irrational to turn down the trade at node 3. Alternatively, we could rely on Minimal Unidimensional Precaution, which also entails that it's irrational to turn down the trade at node 3 (since B^- is a souring of B and you do not prefer A to B^-). But this conclusion, that it's irrational to turn down the trade at node 3, contradicts our assumption that each of nodes 2–4 can be reached without making any irrational choices.

Assume next, for proof by contradiction, that each of nodes 2 and 3 can be reached without making any irrational choices. Then, at nodes 1–3, you retain your rationality and your trust in your rationality at these nodes. Since we have already shown that it's irrational to go down at (at least) one of nodes 1–3, it must be irrational to go down at node 3. So it's rationally required to accept the trade at node 3. So you would accept the trade at node 3. Then, taking this prediction into account at node 2, the choice at that node is effectively between

A^{--} (accepting the trade) and B (turning it down). So you might choose either to accept or to turn down the trade at node 2, since neither of A^{--} and B is preferred to the other (so neither choice at node 2 is irrational). Taking this prediction into account at node 1 with precautionary backward induction, we find that it's irrational to turn down the trade at node 1. This is so, because one of the rationally allowed outcomes of turning the trade down (A^{--}) is less preferred than the prospect of accepting the trade (A^-) but none of the rationally allowed outcomes of turning the trade down (that is, neither A^{--} nor B) is preferred to the prospect of accepting the trade (A^-). Alternatively, we could rely on Minimal Unidimensional Precaution, which also entails that it's irrational to turn down the trade at node 1 (since A^{--} is a souring of A^- and you do not prefer B to A^{--}). But this conclusion, that it's irrational to go down at node 1, contradicts our assumption that each of nodes 2 and 3 can be reached without making any irrational choices.

Finally, assume, for proof by contradiction, that node 2 can be reached without making any irrational choices. Then, at nodes 1 and 2, you retain your rationality and your trust in your rationality at these nodes. Since we have already shown that it's irrational to go down at (at least) one of nodes 1 and 2, it must be irrational to go down at node 2. So you would accept the trade at node 2. Taking this prediction into account at node 1, we find that the choice at that node is effectively between A^- (accepting the trade) or A^{--} (turning it down). Since you prefer A^- to A^{--}, it is irrational to turn the trade down and go to node 2. But this contradicts our assumption that node 2 can be reached without making any irrational choices. Hence it is irrational to go down at node 1. So it's rationally required to accept the trade at node 1. So you accept the trade from A to A^- at node 1. And then you end up with A^- even though you could have kept A for free. And we managed to show this without assuming that you would make rational choices at nodes that can only be reached by irrational choices.

So we have a money-pump argument that Completeness is a requirement of rationality, and this argument relies on the following requirements of rationality:

- Backward induction at nodes that can be reached without making irrational choices
- The Principle of Unexploitability
- Symmetry of Souring Sensitivity

And, in addition, the argument relies on the following principles:

- Decision-Tree Separability

- The Irrationality of Single Sourings
- The possibility of the Precaution Money Pump
- The Principle of Future-Choice Independence
- The Principle of Preferential Invulnerability
- The Principle of Rational Decomposition
- Weak Insensitivity to Souring

Moreover, since we only relied on the Irrationality of Single Sourings, the Principle of Future-Choice Independence, and the Principle of Rational Decomposition in the derivation of Minimal Unidimensional Precaution, we could drop these assumptions if we assume Minimal Unidimensional Precaution as a requirement of rationality. Likewise, since Minimal Unidimensional Precaution as a requirement of rationality is equivalent (given standard backward induction) to precautionary backward induction in the Precaution Money Pump, we may also drop Minimal Unidimensional Precaution in this argument if we assume precautionary backward induction at nodes that can be reached without making irrational choices.

If we also assume that the preferences in (13) are robust for small differences in money, we can create a ruinous version of the Precaution Money Pump by iteratively extending the scheme backwards so that you have an initial choice between (i) paying the exploiter a large sum of money to go away and (ii) not paying the exploiter and then face a long sequence of iterations of the Precaution Money Pump with gradually smaller payments.

If Completeness is a requirement of rationality, does this mean that you are required to form an opinion about all possible pairs of options?[99] This depends on what, more precisely, we take preference relations to be. A preference relation between X and Y is, I suggest, a disposition to have a certain psychologically real mental ranking of X and Y *if you were to compare X and Y.*[100] So you do not have to have a psychologically real mental ranking of all possible options, which would be impossible for our limited minds. It's sufficient to have a psychologically real algorithm for arriving at a mental ranking of the options in case you were to compare them. A calculator may be a helpful analogy. A calculator does not have all sums stored for all pairs of numbers it may be asked to add up. What it has is a (storage-wise) relatively small algorithm for how to calculate these sums when needed. Completeness, given the suggested

---

[99] Anand 1987, pp. 191–2.

[100] This suggestion fits with von Neumann and Morgenstern's (1944, p. 17) original conception of Completeness, which merely requires that, for any two alternatives put before the agent, the agent is able to tell what their preference is between the two.

understanding of preference relations, doesn't demand anything less feasible than that.[101]

4 Transitivity

4.1 Transitivity

Consider, once more, having a cup of coffee with one, two, or three lumps of sugar. Suppose, as before, that you can taste the difference between a cup with one lump and a cup with three lumps but you can neither taste the difference between a cup with one lump and a cup with two lumps nor taste the difference between a cup with two lumps and a cup with three lumps. This time, however, you only care about taste. Accordingly, you're indifferent between having a cup with one lump and having a cup with two lumps and indifferent between having a cup with two lumps and having a cup with three lumps. And, due to your sweet tooth, you prefer having a cup with three lumps to having a cup with one lump.[102]

While your preferences are acyclic, they still violate Transitivity – the second basic axiom of Expected Utility Theory:[103]

Transitivity If $X \succsim Y \succsim Z$, then $X \succsim Z$.

Since we have already established that Acyclicity and Completeness are requirements of rationality, we have already established the irrationality of the following kinds of violations of Transitivity:

(1) $A \succ B \succ C \succ A$.
(14) $A \succsim B \succsim C \parallel A$.

Preferences of the kind in (1) violate Acyclicity, and preferences of the kind in (14) violate Completeness. But these aren't the only kinds of non-transitive preferences. In order to complete the argument that Transitivity is a requirement of rationality, we also need to show that the following complete and acyclic kinds of non-transitive preferences are irrational:[104]

(15) $A \succ B \succ C \sim A$.
(16) $A \succ B \sim C \sim A$.

---

[101] Wakker (2010, p. 38) objects that Completeness requires that we have preferences between prospects that we couldn't possibly choose between. We can handle this worry the same way we handled other non-practical preferences in Section 2.2.

[102] Luce 1956, p. 179.

[103] Arrow 1951, p. 13 and Jensen 1967, p. 171. More generally, a relation over a set is *transitive* if and only if – for all x, y, and z in the set – if x is related to y and y is related to z, then x is related to z. See De Morgan 1851, p. 104.

[104] Nozick 1963, p. 88. This key step is often ignored. See, for instance, Tversky 1969, p. 45.

The money pumps we discussed in Section 2 don't work for agents who have the preferences in (15) or (16). In the Upfront Money Pump, for example, agents with these preferences do not prefer C to A, so they aren't rationally required to accept the trade from A to C. And then the argument falls apart.

This is easiest to see in case of the preferences in (16). Given the prediction that either of A and C may be chosen at node 3, agents with the preferences in (16) can rely on the following dominance version of backward induction – which, given Completeness, is equivalent to precautionary backward induction:

> According to *dominance backward induction*, it is irrational to choose an option X over an option Y if there is a rationally allowed outcome of X (that is, a prospect of an available plan consisting in choosing X followed by choices that are not irrational) that is less preferred than some rationally allowed outcome of Y and every rationally allowed outcome of Y is at least as preferred as every rationally allowed outcome of X.

Dominance backward induction faces much the same worry as Minimal Unidimensional Precaution and precautionary backward induction, namely, that it seems less plausible in case you're sure that, if you were to choose X, you wouldn't make a sequence of choices that has a less preferred prospect than any rationally allowed outcome of Y. But much the same response applies here too (see Section 3).

Given dominance backward induction, the agent with the preferences in (16) is rationally required to turn down the trade at node 2, because the rationally allowed outcomes of doing so (that is, A and C) are both at least as good as the rationally allowed outcome of accepting the trade (B) and one of the rationally allowed outcomes of turning the trade down (A) is preferred to the rationally allowed outcome of accepting it. Plausibly, C is not only at least as preferred as A but also at least as preferred as A^-. Therefore, taking into account at node 1 what may be chosen at later nodes, we find (given dominance backward induction) that it's rationally required to turn down the trade at node 1. Hence the earlier money-pump argument is blocked.

One way to extend the money-pump argument for Acyclicity so that it also works for agents who have the preferences in (15) or (16) is the *souring approach*, which is to convert those acyclic, non-transitive preferences into cyclic preferences of the kind in (1) by breaking the indifferences with the help of sourings. [105]

Suppose that you have the preferences in (15). From (15), we have, by Unidimensional Continuity of Preference,

---

[105] McClennen 1990, pp. 90–1.

(17) $A > A^- > B$,

where A^- is a souring of A.

Now, consider the following requirement of rationality:[106]

> *Unidimensional IP-Transitivity* If (i) Y^- is a souring of Y and (ii) $X \sim Y > Y^-$, then $X > Y^-$.

This requirement is plausible. Since a souring of Y does not improve Y in any dimension the agent cares about, a souring of Y should tip the scale between the two indifferent prospects X and Y in favour of the unsoured X. Nevertheless, Unidimensional IP-Transitivity may seem to assume, to some extent, the point at issue in an argument that Transitivity is a requirement of rationality. We will deal with this worry shortly.

From (15) and (17), we have, by Unidimensional IP-Transitivity,

(18) $C > A^-$.

Then – from (15), (17), and (18) – we have

(19) $B > C > A^- > B$.

We have derived cyclic preferences of the kind in (1), which can be shown to be irrational with the Upfront Money Pump.

Next, suppose that you have the preferences in (16). The first two steps proceed as before. From (16), we have, by Unidimensional Continuity of Preference,

(20) $A > A^- > B$,

where A^- is a souring of A. From (16) and (20), we have, by Unidimensional IP-Transitivity,

(21) $C > A^-$.

From (21), we have, by Unidimensional Continuity of Preference,

(22) $C > C^- > A^-$,

where C^- is a souring of C. From (16) and (22), we have, by Unidimensional IP-Transitivity,

(23) $B > C^-$.

Finally – from (20), (22), and (23) – we have

(24) $B > C^- > A^- > B$.

---

[106] For standard IP-Transitivity, see Halldén 1957, p. 62 and Sonnenschein 1965, p. 625.

As before, we have derived cyclic preferences of the kind in (1), which can be shown to be irrational with the Upfront Money Pump.

Hence we have a money-pump argument that Transitivity is a requirement of rationality, and this argument relies on the following requirements of rationality:

- Backward induction at nodes that can be reached without making irrational choices
- Completeness (defended in Section 3)
- The Principle of Unexploitability
- Unidimensional Continuity of Preference
- Unidimensional IP-Transitivity

And, in addition, the argument relies on the following principles:

- Decision-Tree Separability
- The possibility of the Upfront Money Pump
- The Principle of Preferential Invulnerability

Still, as mentioned, Unidimensional IP-Transitivity is a special case of Transitivity. So it assumes, at least in part, the point at issue in an argument for Transitivity. Many of the alleged counter-examples to Transitivity would also be counter-examples to Unidimensional IP-Transitivity. For instance, let A be a free trip to Austin, let B be a free trip to Boston, and let B^- be the trip to Boston at a cost of \$1.[107] It may seem rationally permitted to be indifferent between A and B and between A and B^- while one prefers B to B^-. If we rebut alleged counter-examples of this kind with the help of Unidimensional IP-Transitivity, we assume the point at issue.[108]

But there is a better response to these alleged counter-examples. If your indifference between two prospects is insensitive to sourings in this manner, then your indifference would have the same problematic insensitivity to sourings as preferential gaps. And then you would be open to a variation of the Precaution Money Pump (from Section 3), replacing preferential gaps with indifference.

Even so, there are other ways to amend the money-pump argument for Transitivity. The following *eventwise approach* makes use of dominance in terms of events. That is, we assume the following requirement of rationality:[109]

---

[107] Restle 1961, pp. 62–3.

[108] Gustafsson 2010, pp. 253–4.

[109] Arrow 1965, p. 17. Savage (1954, pp. 21–2) puts forward the less general *Sure-Thing Principle*, which only differs from the Strong Principle of Eventwise Dominance in that the set of events has to have exactly two elements in the antecedent.

The Strong Principle of Eventwise Dominance If there is a set of events such that (i) the set is a partition of states of nature, (ii), given each event E in the set, the outcome of gamble G given E is at least as preferred as the outcome of gamble G^* given E, and (iii), in some positive-probability event E^* in the set, the outcome of G given E^* is preferred to the outcome of G^* given E^*, then $G > G^*$.

This principle avoids the earlier worry about IP-Transitivity. Note that the standard, alleged, counter-examples to Transitivity would not be counter-examples to the Strong Principle of Eventwise Dominance. So the Strong Principle of Eventwise Dominance does not assume the point at issue against these alleged counter-examples. So it does not assume the point at issue in an argument for Transitivity.

(Yet, since we will rely on Transitivity for the argument for the strong strict-preference version of Independence in Section 5.3, it may seem that the Strong Principle of Eventwise Dominance assumes, in part, the point at issue in an argument for Independence. In this respect, the souring approach is better. Still, in Section 5.1, we will rebut the commonly claimed counter-examples to the Strong Principle of Eventwise Dominance without relying on Transitivity.)

Suppose that you violate Transitivity by having the preferences in one of (1), (15), and (16). Then you have the following preferences:

(25) $A > B \gtrsim C \gtrsim A$.

Now, consider gambles G_1, G_2, and G_3, which have different outcomes in positive-probability events E_1, E_2, and E_3 that are such that $\{E_1, E_2, E_3\}$ is a partition of states of nature:

	E_1	E_2	E_3
G_1	A	B	C
G_2	B	C	A
G_3	C	A	B

From (25), we have, by the Strong Principle of Eventwise Dominance, the following preferences over the gambles:[110]

---

[110] If we restrict Transitivity to sure prospects, then prospects A, B, and C in the violating preferences must be sure prospects. And then we can rely on the following statewise principle (rather than the Strong Principle of Eventwise Dominance):

The Strong Principle of Statewise Dominance If (i), in each state of nature, the sure prospect of the final outcome of gamble G is at least as preferred as the sure prospect of the final outcome of gamble G^* and (ii), in some positive-probability state of nature, the sure prospect of the final outcome of G is preferred to the sure prospect of the final outcome of G^*, then $G > G^*$.

(26) $G_1 > G_2 > G_3 > G_1$.

Once more, we have derived cyclic preferences of the kind in (1), which can be shown to be irrational with the Upfront Money Pump.[111]

So we have a money-pump argument that Transitivity is a requirement of rationality, and this argument relies on the following requirements of rationality:

- Backward induction at nodes that can be reached without making irrational choices
- Completeness (defended in Section 3)
- The Principle of Unexploitability
- The Strong Principle of Eventwise Dominance
- Unidimensional Continuity of Preference

And, moreover, the argument relies on the following principles:

- Decision-Tree Separability
- The possibility of the Upfront Money Pump
- The Principle of Preferential Invulnerability

Hence we have two alternative arguments for Transitivity, which rely on notably different assumptions.

4.2 Transitivity of Strict Preference

While we have already argued that Transitivity is a requirement of rationality, it is worth investigating whether we can make do with more compelling assumptions if we merely seek to defend the following (logically weaker) requirement of rationality:[112]

Transitivity of Strict Preference   If $X > Y > Z$, then $X > Z$.

In order to show that Transitivity of Strict Preference is a requirement of rationality, we need to show that all kinds of violating preferences are irrational. Violations of Transitivity of Strict Preference can be of the following kinds:

(1) $A > B > C > A$.
(15) $A > B > C \sim A$.
(27) $A > B > C \parallel A$.

(See Savage 1951, p. 58 and Milnor 1954, p. 55.) Nevertheless, the argument for the strong strict-preference version of Independence (in Section 5.3) needs Transitivity for prospects in general – not just sure prospects.

[111] Gustafsson 2010, pp. 255–6.

[112] von Neumann and Morgenstern 1944, pp. 26–7.

Notably absent from this list of violations are preferences of the following kind:

(16) $A > B \sim C \sim A$.

The preferences in (16) violate Transitivity (that is, transitivity of *at least as preferred as*) but not Transitivity of Strict Preference. This, as we shall see, allows us to make do without not only Unidimensional IP-Transitivity but also the Strong Principle of Eventwise Dominance.

The preferences in (1) violate Acyclicity, and the preferences in (27) violate Completeness. So, given Acyclicity and Completeness, we only need to show that preferences of the kind in (15) are irrational.

Suppose that you have the preferences in (15). From (15), we have, by Unidimensional Continuity of Preference,

(17) $A > A^- > B$,

where A^- is a souring of A.

We also assume, as a requirement of rationality, the mirror image of Unidimensional Continuity of Preference:

> *Unidimensional Continuity of Dispreference* If $X > Y$, then there is a prospect Y^+ such that (i) Y^+ is a sweetening of Y and (ii) $X > Y^+ > Y$.

The underlying idea behind this principle is the same as for Unidimensional Continuity of Preference: if you (strictly) prefer X to Y, then you must prefer X with some margin. So there should be some, perhaps minimal, amount you are willing to forgo to get X rather than Y.

From (15), we have, by Unidimensional Continuity of Dispreference,

(28) $B > C^+ > C$,

where C^+ is a sweetening of C.

Next, instead of Unidimensional IP-Transitivity, we assume the following requirement of rationality:[113]

> *Unidimensional PI-Acyclicity* If (i) X^+ is a sweetening of X and (ii) $X^+ > X \sim Y$, then it is not the case that $Y > X^+$.

Back in Section 4.1, when we assumed Unidimensional IP-Transitivity for the souring approach, we assumed that the souring of one of two indifferent prospects made it less preferred than the other. And we had to rule out that you may still be indifferent between the prospects. Here, we needn't do so. All we assume is that a sweetening of one of two indifferent prospects does not make it *less* preferred than the other.

---

[113] This is a unidimensional, preferential variant of a principle in Chisholm and Sosa 1966, p. 249.

From (15) and (28), we have, by Unidimensional PI-Acyclicity,

(29) It is not the case that $A > C^+$.

From (29), we have, by Completeness,

(30) $C^+ \gtrsim A$.

Now, consider the decision problem in Figure 12, the *Strict-Preference Money Pump*.

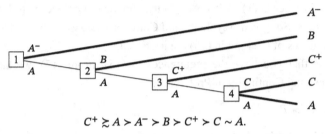

$$C^+ \gtrsim A > A^- > B > C^+ > C \sim A.$$

Figure 12 The Strict-Preference Money Pump

At node 4, you are both rationally permitted to go up and rationally permitted to go down, since you are indifferent between A and C.

Taking this into account at node 3, we find that the prospect of going up (C^+) is preferred to one of the rationally allowed outcomes of going down (C) and the prospect of going up is at least as preferred as every rationally allowed outcome of going down. So, by dominance backward induction, it is rationally required that you accept the trade at node 3. Alternatively, we could rely on Minimal Unidimensional Precaution, which also prescribes going up at node 3 – since C is a souring of C^+ and you do not prefer A to C.

Taking this into account at node 2, it is rationally required to accept the trade at that node, since you prefer B to C^+.

Finally, taking this prediction into account at node 1, it is rationally required to accept the initial trade, since you prefer A^- to B. So you end up with A^- even though you could have kept A for free.

Note that the Strict-Transitivity Money Pump is BI-terminating. So we only need to assume that, at nodes that can be reached without making irrational choices, you retain (i) your rationality and (ii) your trust in your rationality at nodes that can be reached without making irrational choices.

Hence we have a money-pump argument that Transitivity of Strict Preference is a requirement of rationality, and this argument relies on the following requirements of rationality:

- Acyclicity (defended in Section 2.2)
- Backward induction at nodes that can be reached without making irrational choices

- Completeness (defended in Section 3)
- The Principle of Unexploitability
- Unidimensional Continuity of Dispreference
- Unidimensional Continuity of Preference
- Unidimensional PI-Acyclicity

And the argument also relies on the following principles:

- Decision-Tree Separability
- The Irrationality of Single Sourings
- The possibility of the Strict-Preference Money Pump
- The Principle of Future-Choice Independence
- The Principle of Preferential Invulnerability
- The Principle of Rational Decomposition

We only need the Irrationality of Single Sourings, the Principle of Future-Choice Independence, and the Principle of Rational Decomposition to derive Minimal Unidimensional Precaution (see Section 3). So we could drop these assumptions if we assume Minimal Unidimensional Precaution as a requirement of rationality. As a requirement of rationality, Minimal Unidimensional Precaution is equivalent (given standard backward induction) to dominance backward induction in the Strict-Preference Money Pump. So we may also drop Minimal Unidimensional Precaution in this argument if we assume dominance backward induction at nodes that can be reached without making irrational choices.

Note that this approach, which makes do with Unidimensional PI-Acyclicity without Unidimensional IP-Transitivity and the Strong Principle of Eventwise Dominance, does not work for the preferences in (16), where C is indifferent not only to A but also to B. If we tried this approach on those preferences, we would see that, once we have sweetened C, the sweetening of C may be preferred not only to A but also to B. And then the approach is blocked.

4.3 Strong Acyclicity

We can extend both the souring approach and the eventwise approach to cover violations of the following weakening of Transitivity (and strengthening of Acyclicity):[114]

Strong Acyclicity If $X_1 \succsim X_2 \succsim \ldots \succsim X_n$, then it is not the case that $X_n \succ X_1$.

---

[114] Suzumura 1976, p. 387.

Violations of this weaker requirement can only be of the following general kind:

(31) $A_1 > A_2 \gtrsim \ldots \gtrsim A_n \gtrsim A_1$.

So suppose that you violate Strong Acyclicity by having the preferences in (31).

For the souring approach, consider first the case where your preferences are, more specifically, of the following cyclical kind:

(7) $A_1 > A_2 > \ldots > A_n > A_1$.

In that case, we can show that your preferences are irrational with the Upfront Acyclicity Money Pump.

Consider next the remaining case where your preferences over $A_1, A_2, \ldots,$ and A_n do not contain a cycle of strict preference. That is, your preferences are merely *weakly cyclical* – that is, you have a cycle of strict preference except that some (but not all) strict-preference relations in the cycle have been replaced by indifference. Find three prospects – A_i, A_j, and A_k – that are adjacent in this weak cycle such that

(32) $A_i \sim A_j > A_k$.

From (32), we have, by Unidimensional Continuity of Preference,

(33) $A_j > A_j^- > A_k$,

where A_j^- is a souring of A_j. Then, from (32) and (33), we have, by Unidimensional IP-Transitivity,

(34) $A_i > A_j^- > A_k$.

Next, we replace the (32) part of the weak cycle with (34). We repeat this procedure if necessary until we end up with a cycle of strict preference. Finally, we use the Upfront Acyclicity Money Pump to show that these cyclic preferences are irrational.

Hence we have a money-pump argument that Strong Acyclicity is a requirement of rationality, and this argument relies on the following requirements of rationality:

- Backward induction at nodes that can be reached without making irrational choices
- The Principle of Unexploitability
- Unidimensional Continuity of Preference
- Unidimensional IP-Transitivity

And, moreover, the argument relies on the following principles:

- Decision-Tree Separability
- The possibility of the Upfront Acyclicity Money Pump
- The Principle of Preferential Invulnerability

For the eventwise approach, consider gambles G_1, G_2, ... , and G_n, which have different outcomes in positive-probability events E_1, E_2, ... , and E_n that are such that $\{E_1, E_2, ..., E_n\}$ is a partition of states of nature:

$$
\begin{array}{ccccc}
 & E_1 & E_2 & \cdots & E_n \\
G_1 & A_1 & A_2 & \cdots & A_n \\
G_2 & A_2 & A_3 & \cdots & A_1 \\
\vdots & \vdots & \vdots & & \vdots \\
G_n & A_n & A_1 & \cdots & A_{n-1}
\end{array}
$$

From (31), we have, by the Strong Principle of Eventwise Dominance,

(35) $G_1 \succ G_2 \succ \ldots \succ G_n \succ G_1$.

We have derived cyclic preferences of the kind in (7), which can be shown to be irrational with the Upfront Acyclicity Money Pump.

So we have a money-pump argument that Strong Acyclicity is a requirement of rationality, and this argument relies on the following requirements of rationality:

- Backward induction at nodes that can be reached without making irrational choices
- The Principle of Unexploitability
- The Strong Principle of Eventwise Dominance
- Unidimensional Continuity of Preference

And, in addition, the argument relies on the following principles:

- Decision-Tree Separability
- The possibility of the Upfront Acyclicity Money Pump
- The Principle of Preferential Invulnerability

Notably, unlike the money-pump argument for Transitivity, these money-pump arguments for Strong Acyclicity do not rely on Completeness.[115]

---

[115] Bossert, Sprumont, and Suzumura (2005, pp. 186–7) claim that Strong Acyclicity is just what is needed to avoid money pumps. But – as we saw in Section 3 and shall see in Section 5 – there are money pumps for strongly acyclic preferences that violate Completeness or Independence.

5 Independence

Suppose that you prefer $1M ($1,000,000) for sure to a one-in-two chance of
$3M, because you don't want to risk getting nothing. But you also prefer a one-
in-three chance of $3M to a two-in-three chance of $1M, because now there is a
risk of getting nothing with either lottery and you prefer the prize of the former
lottery to the prize of the latter one.[116]

Your preferences violate Independence – which, in one version, is the third
basic axiom of Expected Utility Theory. Let XpY be a prospect consisting in a
lottery between X and Y such that X occurs with probability p and Y occurs with
probability $1 - p$, where X and Y are also prospects that are either lotteries themselves or sure prospects.[117] The most straightforward version of Independence
can then be stated as follows:[118]

> *Independence (the biconditional weak-preference version)*   For all prob-
> abilities p such that $0 < p < 1$, it holds that $X \succsim Y$ if and only if $XpZ \succsim$
> YpZ.

Roughly, the idea is that your preference between two prospects should be the
same if the same chance of a third prospect was added to both. Still, the standard
axiomatization of Expected Utility Theory makes do with a logically weaker
version of Independence. The third basic axiom of Expected Utility Theory is
the following principle:[119]

> *Independence (the strong strict-preference version)*   For all probabilities p
> such that $0 < p < 1$, it holds that, if $X \succ Y$, then $XpZ \succ YpZ$.

A challenge to the idea that Independence is a requirement of rationality is that
the most straightforward version – the biconditional weak-preference version –
conflicts with some seemingly rational preferences, namely, Allais and Ellsberg
Preferences. Those preferences, however, can be shown to be irrational with the
help of a money-pump argument with fairly weak assumptions (Section 5.1).

Furthermore, there is a money-pump argument, with even weaker assumptions, that the following version of Independence is a requirement of rationality (Section 5.2):[120]

---

[116] This is a simplified version of Kahneman and Tversky's (1979, pp. 266–7) variation of the
Allais Paradox (see Section 5.1).
[117] Rabinowicz 1995, p. 588.
[118] Rubin 1949, p. 2. For a historical account of Independence, see Fishburn and Wakker 1995.
[119] Jensen 1967, p. 173.
[120] Gustafsson 2021, p. 23.

Independence (the weak strict-preference version) For all probabilities p such that $0 < p < 1$, it holds that, if $X > Y$, then it is not the case that $YpZ > XpZ$.

But this version of Independence is too weak to characterize Expected Utility Theory together with Completeness, Continuity, and Transitivity. Still, given somewhat stronger assumptions, the money-pump argument for the weak strict-preference version can be extended so that it also works for the strong strict-preference version (Section 5.3). And, with only slightly stronger assumptions, we can show that the biconditional weak-preference version of Independence is a requirement of rationality (Section 5.4).

5.1 Allais, Ellsberg, and Independence for Constant Outcomes

The two most prominent objections to Independence are the Allais Paradox (put forward by Maurice Allais) and the Ellsberg Paradox (put forward by Daniel Ellsberg). These paradoxes directly challenge not only the biconditional weak-preference version of Independence but also the following, logically weaker, requirement: [121]

Independence for Constant Outcomes (the weak strict-preference version) For all probabilities p such that $0 < p < 1$, it holds that, if $XpU > YpU$, then it is not the case that $YpV > XpV$.

Violations of this variant of Independence can only be of the following kind:

(36) $ApC > BpC$, and $BpD > ApD$,

where p is a probability such that $0 < p < 1$. As we shall see, the Allais Paradox and the Ellsberg Paradox both feature seemingly rational preferences of this kind.

The Allais Paradox involves four lotteries. In lottery L_1, one gets \$1M for certain; in lottery L_2, there is a 10% probability of getting \$5M, an 89% probability of getting \$1M, and a 1% probability of getting \$0; in lottery L_3, there is an 11% probability of getting \$1M and an 89% probability of getting \$0; and, in lottery L_4, there is a 10% probability of getting \$5M and a 90% probability of getting \$0: [122]

---

[121] McClennen 1990, pp. xii, 45. In Gustafsson 2021, p. 23, I called this principle 'Independence for Constant Prospects'. But this is potentially confusing, since a 'constant prospect' sometimes means a sure prospect. See, for instance, Wakker 2010, p. 14.

[122] Allais 1953, p. 527; 1979, p. 89. In Allais's version, the prizes are 100 million and 500 million francs. An earlier, similarly structured example was put forward by Morlat (1953, pp. 156–7) at a conference in 1952. A notable difference between Allais's and Morlat's examples is that the

	1%	10%	89%
L_1	$1M	$1M	$1M
L_2	$0	$5M	$1M
L_3	$1M	$1M	$0
L_4	$0	$5M	$0

Many people have the following preferences, which we can call *Allais Preferences*:

(37) $L_1 > L_2$, and $L_4 > L_3$.

To see that Allais Preferences violate the weak strict-preference version of Independence for Constant Outcomes, let A be a sure prospect of $1M; let B be the prospect of a 10/11 probability of $5M, otherwise $0; let C be a sure prospect of $1M (just like A); and let D be a sure prospect of $0:

	1/11	10/11
A	$1M	$1M
B	$0	$5M
C	$1M	$1M
D	$0	$0

If we let p be 11/100, then L_1 is equivalent to ApC, L_2 is equivalent to BpC, L_3 is equivalent to ApD, and L_4 is equivalent to BpD. So (37) can be stated as follows:

(36) $ApC > BpC$, and $BpD > ApD$.

Accordingly, Allais Preferences violate the weak strict-preference version of Independence for Constant Outcomes.

   The Ellsberg Paradox features an urn that is known to contain 30 red balls and 60 balls that are either black or yellow (and this is all that is known with respect to the proportions in the urn). The proportion of black to yellow balls is unknown. A ball will be drawn at random from the urn. Just like the Allais Paradox, the Ellsberg Paradox involves four lotteries. Lottery L_1 pays $100 if the ball is red, otherwise $0; lottery L_2 pays $100 if the ball is black, otherwise $0; lottery L_3 pays $100 if the ball is red or yellow, otherwise $0; and lottery L_4 pays $100 if the ball is black or yellow, otherwise $0:[123]

---

latter doesn't rely on certainty. See Mongin 1999, p. 456 for an account of Morlat's example in English.

[123] Ellsberg 1961, pp. 653–4.

	30 balls	60 balls	
	Red	Black	Yellow
L_1	$100	$0	$0
L_2	$0	$100	$0
L_3	$100	$0	$100
L_4	$0	$100	$100

Many people have the following preferences, which we can call *Ellsberg Preferences*:

(38) $L_1 > L_2$, and $L_4 > L_3$.

Ellsberg Preferences violate the weak strict-preference version of Independence for Constant Outcomes. To see this, let p be the unknown probability of *the ball's being either red or black* (which, given the agent's knowledge, is equivalent to *the ball's not being yellow*); let A be the prospect of a $1/(3p)$ probability of $100, otherwise $0; let B be the prospect of a $1 - 1/(3p)$ probability of $100, otherwise $0; let C be the sure prospect of $0; and let D be the sure prospect of $100:

	$1/(3p)$	$1 - 1/(3p)$
A	$100	$0
B	$0	$100
C	$0	$0
D	$100	$100

We see that L_1 is equivalent to ApC, L_2 is equivalent to BpC, L_3 is equivalent to ApD, and L_4 is equivalent to BpD. So (38) can be stated as follows:

(36) $ApC > BpC$, and $BpD > ApD$.

Accordingly – just like Allais Preferences – Ellsberg Preferences violate the weak strict-preference version of Independence for Constant Outcomes.

Hence both Allais and Ellsberg Preferences entail preferences of the kind in (36), so they both violate the weak strict-preference version of Independence for Constant Outcomes. Consequently, Allais and Ellsberg Preferences violate the (logically stronger) biconditional weak-preference version of Independence.[124]

---

[124] Neither Allais nor Ellsberg Preferences violate the strong strict-preference version of Independence. Still, if we assume that – in addition to having the preferences in (36) – you also prefer one of A and B to the other, then we do get a violation of not only the strong but also the weak strict-preference version of Independence.[125] Nevertheless, having Allais or Ellsberg Preferences does not commit you to having this additional preference.

As we have seen, both Allais and Ellsberg Preferences violate the weak strict-preference version of Independence for Constant Outcomes. Accordingly, Allais and Ellsberg Preferences are both irrational if this principle is a requirement of rationality. Can we show that it is a requirement of rationality? We can.

Suppose that you violate the weak strict-preference version of Independence for Constant Outcomes by having the preferences in (36). From (36), we have, by Unidimensional Continuity of Preference,

(39) $ApC > A^-pC^- > A^{--}pC^{--} > BpC$, and
$\qquad BpD > B^-pD^- > B^{--}pD^{--} > ApD$,

where A^-pC^- and B^-pD^- are sourings of ApC and BpD respectively and where $A^{--}pC^{--}$ and $B^{--}pD^{--}$ are sourings of A^-pC^- and B^-pD^- respectively.

For simplicity, we may assume the following requirement of rationality – even though, strictly, we don't need it:

> *The Souring Principle*   If X^- is a souring of X, then $X > X^-$.

We have, by the Souring Principle,

(40) $A > A^- > A^{--}, B > B^- > B^{--}$, and $C > C^- > C^{--}$.

Now, suppose that E_1 and E_2 are two independent chance events such that E_1 occurs with probability $1/2$ and E_2 occurs with probability p. And consider gambles G_1, G_1^-, and G_2 whose outcomes depend on these two events:

	E_1 & E_2	E_1 & $\neg E_2$	$\neg E_1$ & E_2	$\neg E_1$ & $\neg E_2$
	$(1/2)(p)$	$(1/2)(1-p)$	$(1/2)(p)$	$(1/2)(1-p)$
G_1	A	D	B	C
G_1^-	A^-	D^-	B^-	C^-
G_2	B^{--}	D^{--}	A^{--}	C^{--}

Like before, we assume that the Weak Principle of Equiprobable Unidimensional Dominance is a requirement of rationality. The Weak Principle of Equiprobable Unidimensional Dominance should be acceptable even if one is risk-averse.[126] In terms of risk, the dominated prospect must be less preferable than the dominating prospect. For every potential undesired final outcome of the dominating prospect, the dominated prospect has a corresponding (soured) final outcome with the same probability which is even less preferred. The

---

[126] Buchak (2013, pp. 37–8), who defends Allais Preferences, accepts the Strong Principle of Stochastic Dominance, which is a stronger requirement than the Weak Principle of Equiprobable Unidimensional Dominance.

probability of getting an undesired final outcome must be at least as high in the dominated prospect as in the dominating prospect. In any compelling violation of Independence for Constant Outcomes, no individual preference between two prospects violates the Weak Principle of Equiprobable Unidimensional Dominance. For instance, the Weak Principle of Equiprobable Unidimensional Dominance does not assume the point at issue against Allais and Ellsberg Preferences. None of the pairwise preferences in Allais and Ellsberg Preferences violates the Weak Principle of Equiprobable Unidimensional Dominance.

From (39) or more simply from (40), we have, by the Weak Principle of Equiprobable Unidimensional Dominance,

(41) $G_1 > G_1^- > G_2$.

Now, consider the decision problem in Figure 13, the *Constant-Outcomes Money Pump*.[127]

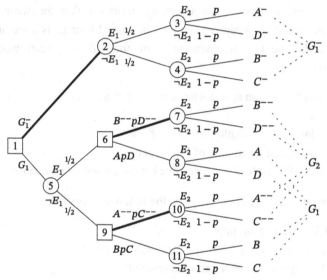

$$G_1 > G_1^- > G_2, A^{--}pC^{--} > BpC, \text{ and } B^{--}pD^{--} > ApD.$$

Figure 13 The Constant-Outcomes Money Pump

At the initial choice node, you have a choice whether to accept a trade from G_1 to G_1^-. If you turn down this trade, chance node 5 determines (depending on E_1) whether you will face node 6 or 9. If E_1 occurs, you are offered a trade

---

[127] This is a variation of an argument in Raiffa 1968, pp. 83–5. See also Raiffa's (1961, p. 694) earlier argument, which uses similar reasoning but doesn't involve any dominance violation. Moreover, unlike this case, Raiffa's two cases are not money pumps, since the agent would not end up with a statewise souring of the walk-away prospect. This is also so for the case in Gustafsson 2021, p. 27. Note also that, unlike the cases in Al-Najjar and Weinstein 2009, pp. 258–66, this case is BI-terminating.

at node 6 from ApD to $B^{--}pD^{--}$. And, if E_1 does not occur, you are offered a trade at node 9 from BpC to $A^{--}pC^{--}$.

At node 6, you would accept the trade from ApD to $B^{--}pD^{--}$, since you prefer $B^{--}pD^{--}$ to ApD. And, at node 9, you would also accept the trade from BpC to $A^{--}pC^{--}$, since you prefer $A^{--}pC^{--}$ to BpC.

Taking this into account at node 1, the choice at that node is effectively between G_1^- (accepting the trade) and G_2 (turning it down). Since you prefer G_1^- to G_2, you accept the initial trade. So you end up with G_1^- when you could have kept G_1 for free. Moreover, note that G_1^- is less preferred than G_1 in every state of nature.

Hence we have a money-pump argument that preferences of the kind in (36) are irrational. Moreover, since the Constant-Outcomes Money Pump is BI-terminating, this argument need not assume that you retain your rationality at nodes that can't be reached without making irrational choices.

Accordingly, we have a money-pump argument that the weak strict-preference version of Independence for Constant Outcomes is a requirement of rationality, and this argument relies on the following requirements of rationality:

- Backward induction at nodes that can be reached without making irrational choices
- The Principle of Unexploitability
- Unidimensional Continuity of Preference
- The Weak Principle of Equiprobable Unidimensional Dominance

And, moreover, the argument relies on the following principles:

- Decision-Tree Separability
- The possibility of the Constant-Outcomes Money Pump
- The Principle of Preferential Invulnerability

Since we can show that the weak strict-preference version of Independence for Constant Outcomes is a requirement of rationality and thereby that Allais and Ellsberg Preferences are irrational, we can rebut the most prominent objections to Independence.

5.2 The weak strict-preference version of Independence

Having rebutted the most prominent objections to Independence (that is, the alleged rationality of Allais and Ellsberg Preferences), let us explore whether there are any compelling *positive* arguments that Independence is a requirement of rationality. We begin with the weakest version of Independence, namely,

Independence (the weak strict-preference version) For all probabilities p such that $0 < p < 1$, it holds that, if $X > Y$, then it is not the case that $YpZ > XpZ$.

This version of Independence can be shown to be a requirement of rationality with the help of a money-pump argument with even weaker assumptions than those we relied on for the argument that Independence for Constant Outcomes is a requirement of rationality.

Violations of the weak strict-preference version can only be of the following kind, where p is a probability such that $0 < p < 1$:

(42) $A > B$, and $BpC > ApC$.

So suppose that you violate the weak strict-preference version of Independence by having the preferences in (42). From (42), we have, by Unidimensional Continuity of Preference,

(43) $BpC > B^-pC^- > ApC$,

where B^-pC^- is a souring of BpC.

Now, consider the decision problem in Figure 14, the *Independence Money Pump*.[128]

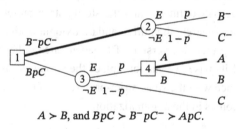

$A > B$, and $BpC > B^-pC^- > ApC$.

Figure 14 The Independence Money Pump

Here, the two chance nodes depend on the same event E, which occurs with probability p. You start off with BpC. At node 1, you are offered a trade from BpC to B^-pC^-. If you accept this trade, then, if E occurs, you end up with B^- and, if E does not occur, you end up with C^-. If you turn the trade down and E occurs, you will be offered a trade from B to A at node 4. And, if you turn down the trade at node 1 and E does not occur, you end up with C.

Since you prefer A to B, you would accept the trade at node 4. Using backward induction at node 1, the prospect of going down is then effectively ApC and the prospect of going up is B^-pC^-. So you go up at node 1, since you prefer B^-pC^- to ApC. But then you end up with B^-pC^- when you could have kept BpC for free if you had followed the plan to go down at each choice node. And,

[128] Hammond 1988b, pp. 292–3, Steele 2010, pp. 469–70, and Gustafsson 2021, p. 31n21.

since the chance nodes depend on the same event, we find that, in every state of nature, the prospect of going up at node 1 is a souring of the prospect of following the plan to go down at each choice node.

Accordingly, we have a money-pump argument that the weak strict-preference version of Independence is a requirement of rationality, and this argument is based on the following requirements of rationality:

- Backward induction at nodes that can be reached without making irrational choices
- The Principle of Unexploitability
- Unidimensional Continuity of Preference

And, in addition, the argument relies on the following principles:

- Decision-Tree Separability
- The possibility of the Independence Money Pump
- The Principle of Preferential Invulnerability

Still, axiomatizations of Expected Utility Theory typically need a stronger version of Independence, like the strong strict-preference version.

As mentioned earlier, Expected Utility Theory can be axiomatized by Completeness, Transitivity, Continuity, and the strong strict-preference version of Independence. Can we strengthen this standard axiomatization so that it relies on the *weak* strict-preference version of Independence rather than the strong one? We cannot. Likewise, we cannot replace the strong strict-preference version of Independence with the weak strict-preference version of Independence for Constant Outcomes in the axiomatization.[129]

5.3 The strong strict-preference version of Independence

So let us turn to

> *Independence (the strong strict-preference version)*   For all probabilities p such that $0 < p < 1$, it holds that, if $X > Y$, then $XpZ > YpZ$.

The good news is that there is a money-pump argument that this version of Independence is a requirement of rationality; the bad news is that this argument requires notably stronger assumptions than the argument for the weak strict-preference version. In order to show that the strong strict-preference version is a requirement of rationality, it's not enough to show that preferences of the following kind are irrational:

(42) $A > B$, and $BpC > ApC$.

---

[129] For proofs of both claims, see Gustafsson 2021, pp. 32–3.

We also need to show the irrationality of violations of the following kinds, where (as before) p is a probability such that $0 < p < 1$:

(44) $A > B$, and $ApC \sim BpC$.
(45) $A > B$, and $ApC \parallel BpC$.

The money-pump argument in Section 5.2 doesn't work against the preferences in (44) and (45). Preferences of the kind in (45) are ruled out by Completeness. The preferences in (44) are more challenging. These preferences violate the strong strict-preference version of Independence, but they do not violate the other standard axioms of Expected Utility Theory.[130] And, since the biconditional weak-preference version of Independence is logically stronger than the strong strict-preference version, the preferences in (44) violate that version too. Hence, to have a cogent argument that these versions of Independence are requirements of rationality, we must show that the preferences of the kind in (44) are irrational.

To establish the irrationality of preferences of the kind in (44), we assume that the following dominance principle is a requirement of rationality:

> *The Strong Principle of Unidimensional Stochastic Dominance*   If (i) X^- is a souring of X, (ii) $X > X^-$, and (iii) $0 < p < 1$, then $XpY > X^-pY$.

Just like the Weak Principle of Equiprobable Unidimensional Dominance, this requirement should be acceptable even if one is risk-averse. The probability of getting an undesired final outcome must be at least as high in the dominated prospect as in the dominating prospect.[131] In any compelling violation of Independence, the individual pairwise preferences do not violate the Strong Principle of Unidimensional Stochastic Dominance.

We will show that preferences of the kind in (42) can be derived from (44) – given that the Strong Principle of Unidimensional Stochastic Dominance, Transitivity, and Unidimensional Continuity of Preference are requirements of rationality.

From (44), we have, by Unidimensional Continuity of Preference,

(46) $A > A^- > B$,

where A^- is a souring of A. From (46), we have, by the Strong Principle of Unidimensional Stochastic Dominance,

(47) $ApC > A^-pC$.

---

[130] Gustafsson 2021, p. 34n23.

[131] Buchak (2013, pp. 37–8), who defends Allais Preferences, accepts the Strong Principle of Stochastic Dominance, which is stronger than the Strong Principle of Unidimensional Stochastic Dominance.

From (44) and (47), we have, by Transitivity,

(48) $BpC > A^-pC$.

Finally, from (46) and (48), we have

(49) $A^- > B$, and $BpC > A^-pC$.

Hence, from (44), we have derived preferences of the kind in (42). Since preferences of that kind can be shown to be irrational by the money-pump argument for the weak strict-preferences version (Section 5.2), we can show that preferences of the kind in (44) are irrational.

Accordingly, we have a money-pump argument that the strong strict-preference version of Independence is a requirement of rationality, and this argument relies on the following requirements of rationality:

- Backward induction at nodes that can be reached without making irrational choices
- Completeness (defended in Section 3)
- The Principle of Unexploitability
- The Strong Principle of Unidimensional Stochastic Dominance
- Transitivity (defended in Section 4.1)
- Unidimensional Continuity of Preference

And, in addition, the argument relies on the following principles:

- Decision-Tree Separability
- The possibility of the Independence Money Pump
- The Principle of Preferential Invulnerability

Note that this argument requires stronger assumptions than those needed in the argument for the weak strict-preference version of Independence, since that argument did not need Completeness, the Strong Principle of Unidimensional Stochastic Dominance, and Transitivity.

5.4 The biconditional weak-preference version of Independence

Finally, let us turn to

> *Independence (the biconditional weak-preference version)*   For all probabilities p such that $0 < p < 1$, it holds that $X \gtrsim Y$ if and only if $XpZ \gtrsim YpZ$.

With only slightly stronger assumptions than those for the argument that the strong strict-preference version is a requirement of rationality, we can show that the biconditional weak-preference version is a requirement of rationality.

In addition to preferences of the kind in (42), (44), and (45) which we have already shown are irrational (with the arguments in Sections 5.2 and 5.3), violations of the biconditional weak-preference version of Independence can also be of the following kinds, where again p is a probability such that $0 < p < 1$:

(50) $A \sim B$, and $ApC > BpC$.
(51) $A \sim B$, and $ApC \parallel BpC$.
(52) $A \parallel B$, and $ApC \gtrsim BpC$.

Two of these violations – namely, (51) and (52) – are ruled out by Completeness. So, to finish the argument that the biconditional weak-preference version is a requirement of rationality, we need only show that preferences of kind in (50) are irrational.

From (50), we have, by Unidimensional Continuity of Preference,

(53) $ApC > A^-pC^- > BpC$,

where A^-pC^- is a souring of ApC. Since A^-pC^- is a souring of ApC, it follows that A^- is a souring of A and that C^- is a souring of C. So we have, by the Souring Principle,

(54) $A > A^-$, and $C > C^-$.

From (50) and (54), we have, by Transitivity,

(55) $B > A^-$.

From (54), we have, by the Strong Principle of Unidimensional Stochastic Dominance,

(56) $A^-pC > A^-pC^-$.

From (53) and (56), we have, by Transitivity,

(57) $A^-pC > BpC$.

Finally, from (55) and (57), we have

(58) $B > A^-$, and $A^-pC > BpC$.

We have, once more, derived preferences of the same kind as those in (42). And, since such preferences can be shown to be irrational by the money-pump argument in Section 5.2, we can show that preferences of the kind in (50) are irrational.

Hence we have a money-pump argument that the biconditional weak-preference version of Independence is a requirement of rationality, and this argument relies on the following requirements of rationality:[132]

- Backward induction at nodes that can be reached without making irrational choices
- Completeness (defended in Section 3)
- The Principle of Unexploitability
- The Souring Principle
- The Strong Principle of Unidimensional Stochastic Dominance
- Transitivity (defended in Section 4.1)
- Unidimensional Continuity of Preference

And the argument also relies on the following principles:

- Decision-Tree Separability
- The possibility of the Independence Money Pump
- The Principle of Preferential Invulnerability

Note that, apart from the addition of the Souring Principle, this argument for the biconditional weak-preference version does not require stronger assumptions than the argument for the strong strict-preference version in Section 5.3.

6 Continuity

Suppose that you prefer having two candy bars to having one candy bar and that you prefer either of these alternatives to suddenly dying. Yet you're not willing to risk any chance of sudden death for a chance of having two candy bars rather than just one.[133]

Your preferences violate the fourth and final basic axiom of Expected Utility Theory, namely, Continuity:[134]

---

[132] This argument also supports that the following, logically weaker, version of Independence is a requirement of rationality:

> *Independence (the strong equal-preference version)*   For all probabilities p such that $0 < p < 1$, it holds that, if $X \sim Y$, then $XpZ \sim YpZ$.

This version was proposed by Marschak (1950, pp. 120–1) and Nash (1950, p. 156). Violations of the strong equal-preference version of Independence can only be of the kinds in (50) and (51).

[133] Alchian 1953, pp. 36–7.

[134] von Neumann and Morgenstern 1944, pp. 26–7, Blackwell and Girshick 1954, p. 106, and Jensen 1967, p. 173. Herstein and Milnor (1953, p. 293) present the following, more mathematically complex, alternative to Continuity:

> *Mixture Continuity*   The sets $\{p \mid XpY \succsim Z\}$ and $\{p \mid Z \succsim XpY\}$ are closed.

Continuity If $X > Y > Z$, then there are probabilities p and q such that (i) $0 < p < 1$, (ii) $0 < q < 1$, and (iii) $XpZ > Y > XqZ$.

Violations of Continuity can be of the following kinds:

(59) $A > B > C$, and $ApC > B$ for all probabilities p such that $0 < p < 1$.

(60) $A > B > C$, and $B > ApC$ for all probabilities p such that $0 < p < 1$.

(61) $A > B > C$, and $B \sim ApC$ for some probability p such that $0 < p < 1$, and

    (i)   there is no probability q such that $0 < q < 1$ and $AqC > B$ or

    (ii) there is no probability r such that $0 < r < 1$ and $B > ArC$.

(62) $A > B > C$, and $B \parallel ApC$ for some probability p such that $0 < p < 1$, and

    (i)   there is no probability q such that $0 < q < 1$ and $AqC > B$ or

    (ii) there is no probability r such that $0 < r < 1$ and $B > ArC$.

Preferences of the kind in (62) are ruled out by Completeness. And preferences of the kind in (61) are ruled out by Transitivity and the strong strict-preference version of Independence. To see this, note that (61) entails that there is a probability $0 < p < 1$ such that

(63) $A > B > C$, and $B \sim ApC$.

From (63), we have, by Transitivity,

(64) $A > C$.

Let q and r be probabilities such that $0 < r < p < q < 1$. Then, from (64), we have, by the strong strict-preference version of Independence,

(65) $AqC > ApC > ArC$.

Finally, from (63) and (65), we have, by Transitivity,

(66) $AqC > B > ArC$.

And (66) rules out (61), since $0 < r < q < 1$.

So, to complete the argument that Continuity is a requirement of rationality, we must show that the remaining kinds of violations are irrational. That is, we need to show that preferences of the kind in (59) and (60) are irrational.

Given both Completeness and Transitivity, Mixture Continuity is a strengthening of Continuity. See Hammond 1998, pp. 155–7.

Suppose that you violate Continuity by having the preferences in (59). From (59), we have, by Unidimensional Continuity of Dispreference,

(67) $B > C^+ > C$,

where C^+ is a sweetening of C such that C^+ is certainly ϵ units superior to C in a dimension you care about. From (59) and (67), we have, by Transitivity,

(68) $ApC > C^+ > C$ for all probabilities p such that $0 < p < 1$.

Now, we will do some relabelling. Let 'A', 'C', and 'C^+' now be 'A^-', 'C^-', and 'C' respectively. Given this relabelling, (68) becomes

(69) $A^-pC^- > C > C^-$ for all probabilities p such that $0 < p < 1$.

We now let A be a sweetening of A^- such that A is certainly ϵ units superior to A^- in the dimension that C and C^- differ. Consider the decision problem in Figure 15, the *Lexi-Optimist Pump*.[135]

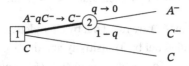

$A^-pC^- > C > C^-$ for all probabilities p such that $0 < p < 1$.

Figure 15 The Lexi-Optimist Pump

Here, no matter how close q gets to 0 (without reaching 0), you still pay the fixed positive amount ϵ to get AqC. So, starting off with C, you are still willing to pay a fixed – not arbitrarily small – amount ϵ to trade C for ApC, which is arbitrarily similar to C (that is, arbitrarily likely to result in the same final outcome as C). So an exploiter can get a payment of ϵ from you with only an arbitrarily small chance of having to give you anything – that is, the arbitrarily small chance of having to trade you A for C. This is arbitrarily close to pure exploitation. You violate the following requirement:

> *The Principle of Limit Unexploitability*   If (i) ϵ is a fixed positive amount, (ii) X^- is a souring of X such that X^- is certainly ϵ units inferior to X in a dimension one cares about, (iii) $X > X^-$, (iv), at node n, P and P^- are two available plans such that P is the only available plan that amounts to walking away from all offers by an exploiter and the prospect of following P is X and the prospect of following P^- is arbitrarily likely to be X^-, and (v) one knows what decision problem one faces at n, then one does not follow P^- from n.

---

[135] This case and the Lexi-Pessimist Pump have a similar structure as Hammond's (1998, pp. 180–1) continuous-behaviour argument for Continuity.

Is it as plausible that the Principle of Limit Unexploitability is a requirement of rationality as that the Principle of Unexploitability is one? Maybe not, but the former is still compelling as a requirement of rationality.[136] If you violate the Principle of Limit Unexploitability, an exploiter can get a fixed amount of money from you for the arbitrarily small cost of an arbitrarily small chance of having to trade you something (in this case, the chance of having to trade you A for C).

It may be objected that, if you prefer A to C by an infinite amount, then it's not a clear sign of irrationality to choose $A^- qC^-$ rather than C with q arbitrarily close to 0. Yet, if you prefer A to C by an infinite amount, you should presumably be willing to pay not only a small amount but any finite amount to get AqC rather than C. So you would pay an arbitrarily large amount to increase the chance of getting A rather than C by an arbitrarily small amount, which seems fanatic.[137]

Next, suppose that you violate Continuity by having the preferences in (60). From (60), we have, by Unidimensional Continuity of Preference,

(70) $A \succ A^- \succ B$,

where A^- is a souring of A which is certainly ϵ units inferior in a dimension you care about. From (60) and (70), we have, by Transitivity,

(71) $A \succ A^- \succ ApC$ for all probabilities p such that $0 < p < 1$.

Finally, consider the decision problem in Figure 16, the *Lexi-Pessimist Pump*.

---

[136] We need the Principle of Limit Unexploitability rather than just the Principle of Unexploitability. Consider

> *Stochastic Leximin* Prospects X and Y are indifferent if and only if, for all possible final outcomes o, the probability of a final outcome that is indifferent to o is the same for X and Y. And X is preferred to Y if and only if there is a possible final outcome o such that (i) the probability of a final outcome that is indifferent to o is greater for X than for Y and (ii), for all possible final outcomes o^* such that o is preferred to o^*, the probability of a final outcome that is indifferent to o^* is the same for X and Y.

Assuming transitive and complete preferences over final outcomes, Stochastic Leximin satisfies Completeness, Transitivity, and Independence. Yet it violates Continuity and cannot be money pumped in a way that leads to a sure loss. See Kowalczyk (2020).

[137] Bostrom (2011, p. 36) calls this the fanaticism problem. Still, infinite differences in preference may be precisely what could justify this kind of fanatic behaviour. Accordingly, the money-pump argument for Continuity seems compelling in case the agent has no (rationally allowed) infinite differences in their preferences.

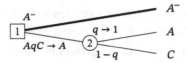

$A > A^- > ApC$ for all probabilities p such that $0 < p < 1$.

Figure 16 The Lexi-Pessimist Pump

In this case, no matter how close q gets to 1 (without reaching 1), you still pay the fixed amount ϵ to get A. So, starting off with AqC, which is arbitrarily likely to result in the same final outcome as A, you are still willing to pay a fixed positive amount ϵ to get A instead. So you violate the Principle of Limit Unexploitability.

Hence we have an argument (which is almost a money-pump argument) that Continuity is a requirement of rationality, and this argument relies on the following requirements of rationality:

- Completeness (defended in Section 3)
- The Principle of Limit Unexploitability
- The strong strict-preference version of Independence (defended in Section 5.3)
- Transitivity (defended in Section 4.1)
- Unidimensional Continuity of Dispreference
- Unidimensional Continuity of Preference

And, in addition, the argument relies on the following principles:

- The possibility of the Lexi-Optimist Pump
- The possibility of the Lexi-Pessimist Pump
- The Principle of Preferential Invulnerability

This argument – when added to the arguments in Sections 3, 4.1, and 5.3 – completes the overall argument that rational preferences conform to Expected Utility Theory.

7 Against Resolute Choice

A common objection to money-pump arguments is that they don't work against agents who follow resolute choice. *Resolute choice* is the approach of choosing in accordance with the plans one has adopted even if one wouldn't choose in accordance with those plans if one hadn't adopted them.[138] In the Upfront Money Pump, for instance, a resolute agent with the cyclic preferences in (1)

---

[138] McClennen 1985, pp. 102–3; 1990, pp. 12–13.

could stick to the plan of turning down all trades so that they end up with A rather than A^-. And then the money-pump argument is blocked.

While resolute choice may seem like a plausible response to money-pump arguments, that plausibility evaporates once it is spelled out how the approach is supposed to work. So how, more precisely, is the resolute-choice approach supposed to work? In the literature, there are six separate ways to be resolute: the Counter-Preferential Approach, the Revision Approach, the Constraint Approach, the Second-Order Approach, the Fine-Grained Approach, and the Conservative Approach. As we shall see, these approaches are all implausible as responses to money-pump arguments.

First, consider the Counter-Preferential Approach:[139]

> *The Counter-Preferential Approach* When you adopt a plan, you follow that plan even if you prefer not to follow it.

On this approach, if you adopt the plan to stick with A in the Upfront Money Pump, you will follow this plan even though you prefer to deviate at node 3. And then you avoid exploitation.

The problem with this approach is that choosing against your own preference at the moment of choice is irrational.[140] The descriptions of the final outcomes should capture everything you care about. So your preferences over these final outcomes and prospects should capture everything you care about.[141] And, if you all-things-considered prefer to deviate from the plan, it would be irrational to follow the plan.

It may be objected that there is still an instrumental rationale for following a plan even though you prefer to deviate – namely, the rationale that the prospect of following the plan at the later node is preferred to what was, at the node the plan was adopted, the prospect of not adopting the plan.[142] But this rationale (which may be compelling at the node you adopt the plan) is no longer compelling at the later node where you prefer to deviate. Because, once you prefer to deviate from the plan, the plan no longer achieves your ends. The alternative to the prospect of following the plan is no longer the prospect you would have faced had you not adopted the plan; the alternative is a prospect you prefer to the prospect of following the plan. So this instrumental rationale for the Counter-Preferential Approach is implausible.[143]

Next, consider the Revision Approach:[144]

[139] Gauthier 1996, p. 220 and Buchak 2013, p. 177.
[140] Or, at least, it seems so if the number of options is finite. See note 164.
[141] Steele 2007, p. 45; 2018, p. 662.
[142] Gauthier 1996, pp. 230–2.
[143] Bratman 1992, pp. 12–13; 1998, pp. 60–1.
[144] McClennen 1985, pp. 102–3; 1990, pp. 213–15.

The Revision Approach   When you adopt a plan, you revise your preferences so that you prefer to act in accordance with the plan at all future choice
nodes (even though you would have preferred to deviate from the plan if you
had not adopted it).

Once you adopt the plan to turn down all offers in the Upfront Money Pump
following the Revision Approach, you prefer A to each of A^-, B, and C. Hence
you would no longer be tempted to accept any of the offers. And then you end
up with A and avoid exploitation.

The Revision Approach has a significant drawback as a defence against
money-pump *arguments* (as opposed to money pumps): If agents with a certain
set of preferences have to adopt some other set of preferences to escape exploitation, then the original preferences still seem irrational. So, in this case, the
defeat of money pumps is a victory for money-pump arguments.[145]

Moreover, following the Revision Approach, there are at least two separate
ways of turning down the first offer in the Upfront Money Pump. You could
turn down the first offer without adopting the plan to turn down all offers (without revising your preferences) and, alternatively, you could turn down the first
offer by adopting that plan (and thereby revise your preferences). But, if the
latter option is available at the first node, it should be added to the decision
tree. (The decision tree should reflect all your choices in the decision problem.) This shows that the Revision Approach does not apply to the original
decision problem where this option is unavailable.

Furthermore, if we include the extra option of adopting the plan to turn down
all trades and thereby revise your preferences, the prospect of that option need
not be the same as the prospect of turning down all trades without having
revised your preferences. So we may include this extra option yet make it unattractive (that is, unattractive before you revise your preferences) by imposing
a cost to revising your preferences. And then, since this option would be less
preferred than the other options given that the cost is sufficiently high, it would
be irrational to choose this extra option. And, once this extra option is ruled
out, you effectively face the original exploitation scheme.

It may be objected that there are many prospects you haven't considered and
you only form your view about them once you face a choice between them.

---

[145] And, since we're defending money-pump arguments rather than money pumps as effective
exploitation schemes, we shouldn't be worried about the empirical finding that people do not
let themselves be continuously money pumped. (See Arkes et al. 2016, p. 23.) In fact, it would
be more worrying for money-pump arguments if otherwise rational people let themselves be
continuously money pumped by refusing to revise their preferences. If you revise your preferences so that you avoid getting money pumped, you seem to have learned the lesson of the
money-pump arguments.

So it may seem rationally permitted to have a preferential gap between some prospects and then revise this gap to a strict preference or indifference once you face a choice between them. This objection, however, relies on an implausible understanding of preferential gaps. The existence of prospects that you still haven't compared does not entail that you have a preferential gap between those prospects. As sketched earlier (at the end of Section 3), I take a preference relation to be a disposition to have a psychologically real mental ranking of the prospects if you were to compare them. You don't need to have a preferential gap between two prospects when you merely haven't got around to comparing them. You have a preferential gap between two prospects in case neither prospect would rank at least as high as the other in your mental ranking if you were to compare them. If preferential gaps in this sense need to be revised in order to avoid money pumps, such gaps are irrational.

Now, consider the Constraint Approach:[146]

> *The Constraint Approach*   When you adopt a plan, you are no longer able to deviate from the plan at future choice nodes.

If you follow the Constraint Approach and adopt the plan to turn down all offers in the Upfront Money Pump, you're no longer able to accept the offers at nodes 2 and 3. So you would end up with A and avoid exploitation.

Nevertheless, the Constraint Approach requires an implausible account of ability.[147] It seems that, after you have adopted a plan, you can still deviate from the plan. Even if you will in fact end up sticking with the plan at later choice nodes, you still have the ability to deviate at those nodes. But this picture, which is suggested by the phenomenology of planning, conflicts with the Constraint Approach.

Furthermore, like the Revision Approach, the Constraint Approach conflicts with the specifications of the decision problems for the money-pump arguments. Consider, for instance, the Upfront Money Pump. In the original decision problem, you do have the opportunity to accept the offers at nodes 2 and 3. So to adopt the plan to turn down all offers in a way that removes those later opportunities would be an additional option at the initial node.[148] But then the Constraint Approach doesn't apply to the original decision problem where this option is unavailable at the initial node. So the Constraint Approach

---

[146] Strotz 1955–6, p. 173 and Hammond 1976, pp. 162–3.

[147] McClennen and Shapiro 1998, p. 367.

[148] Like Homer's (*Od.* 12.50–4; 1995, p. 453) Odysseus, who, *in addition* to sailing past the Sirens without being tied to the mast, has the option of sailing past the Sirens while being tied to the mast.

does not help you escape being money pumped in the original decision problem.[149]

And, as before, we also note that, if we include the extra option of constraining yourself to turning down all trades, the prospect of that option need not be the same as the prospect of turning down all trades without having constrained yourself. So we may include this extra option yet make it unattractive by imposing a cost to constraining yourself. Since this option would be less preferred than the other options (given that the cost is sufficiently high), it would be irrational to choose this extra option. And then, once this extra option is ruled out, you effectively face the original exploitation scheme.

Let us turn to the Second-Order Approach:[150]

> *The Second-Order Approach*   When you adopt a plan, your first-order preferences remain the same but you have a second-order preference for steadfastness which motivates choosing to act in accordance with the plan at all future nodes.

You can follow the Second-Order Approach by having a second-order preference for not being exploited which trumps your other preferences. So, even though you prefer A^- to B in the Standard Money Pump, you would turn down the offer to trade from B to A^- at node 3. What counts against A^- at node 3 is that you could have had A (had you chosen otherwise earlier). So, even though you prefer A^- to B, you prefer B to A^--*when-you-could-have-had-A*.

A problem for the Second-Order Approach is that any second-order preference for honouring previous commitments should be included in your overall, first-order preferences, which are what the axioms of Expected Utility Theory apply to.[151] And then, given that your overall, first-order preferences concern more complex options that include previous commitments, there's no need to introduce any second-order preferences.

This problem, however, does not apply to a first-order variation of the Second-Order Approach:[152]

> *The Fine-Grained Approach*   When you adopt a plan, you prefer to stick to that plan and, since you care about whether you stick to the plan, the final outcomes include information about whether they were reached by sticking to the plan.

More generally, you care not only about what your final holding will be but also about how you arrive at that final holding relative to the whole decision tree.

[149] Hammond 1976, pp. 162–3.
[150] McClennen 1997, p. 239.
[151] Steele 2007, pp. 44–5; 2018, p. 662.
[152] Machina 1989, pp. 1647–9 and McClennen 2009, pp. 133–5.

So your first-order preferences are over prospects where the final outcomes include information about what could have been chosen in the decision tree.[153]

Once we have made sure that your preferences cover prospects with final outcomes that cover everything you care about, it may be that the decision problems we have relied on for the money-pump arguments are no longer possible. Because the final outcomes would include information about what the rest of the decision tree looks like, that information restricts what decision trees they could be part of. Likewise, once we transform the final outcomes in a decision problem so that they include the information about the rest of the decision problem, the agent need no longer have the preferences we have assumed over their options at the choice nodes.

Nevertheless, the Fine-Grained Approach conflicts with some plausible restrictions on what it is rationally permitted to prefer. Suppose that, in the Upfront Money Pump, you adopted at node 1 the plan to stick with *A* (that is, you plan to turn down all trade offers). At node 3, you have a choice between *A* and *C*. Yet, given your fine-grained preference for sticking to your plan, we should represent these options as *A-and-sticking-to-the-plan* and *C-and-deviating-from-the-plan*. Even though you prefer *C* to *A*, you prefer *A-and-sticking-to-the-plan* to *C-and-deviating-from-the-plan*. But why would you care at node 3 whether you stick to the plan you adopted at node 1? Whatever reasons you had for adopting that plan are no longer relevant. The bare fact that you decided to adopt the plan in the past does not provide a reason for sticking to the plan.[154] And, once the fact that you adopted the plan is left out, not only do you now prefer that you choose *C* over *A* but you also preferred, at the time you adopted the plan, that you would choose *C* over *A*. So caring about this plan for its own sake seems irrational. (This substantial restriction on what can be rationally preferred is, of course, a departure from a purely formal approach to decision theory that imposes no such restrictions.[155])

Next, suppose that you have an overriding preference not to be money pumped.[156] Accordingly, in any choice between two options where you are money pumped in one option but not in the other, you prefer not to be money pumped. But caring about being money pumped for its own sake seems irrational. Consider, for instance, the choice at node 3 in the Standard Money Pump. At this node, you have a choice between *A⁻* and *B*. Given your fine-grained preference for not being money pumped, we should represent these

---

[153] Schick 1986, p. 118 and Anand 1993a, pp. 62–4; 1993b, pp. 342–3.

[154] Broome 1992, p. 668.

[155] Broome (1993, pp. 56–9) argues that we need some restrictions on what we can rationally care about or the requirements of rationality would be toothless.

[156] McClennen 2009, p. 134.

options as A^--*and-being-money-pumped* and *B-and-not-being-money-pumped*.
Even though you prefer A^- to B, you prefer *B-and-not-being-money-pumped* to
A^--*and-being-money-pumped*. Choosing A^- at node 3 amounts to being money
pumped because you could have kept A for free if you had turned down the trade
at node 1. But, at node 3, why would you care whether you could have had A
rather than A^-?[157] The loss of the opportunity to have A for free is a sunk cost
at that point.[158] To prefer, for its own sake, that one isn't money pumped seems
irrational.

(Does this claim rule out the irrationality of being money pumped? It does
not. It is irrational to pay for something you could keep for free given that you
prefer having more money other things being equal, but it needn't be irrational
to pay for something you can no longer have for free.)

It may be objected that, in the Upfront Money Pump (and other BI-
terminating money pumps), the Fine-Grained Approach need not attach any
significance to sunken costs. To block exploitation, the fine-grained preference
only needs to make you turn down the first trade. There is no sunk cost when
you consider the first trade, since you are still at the initial node.

But consider the choice nodes you could face if you were to turn down
the trade at the initial node, that is, nodes 2 and 3. At those nodes, you can
no longer be money pumped. So, for the feasible prospects that remain at
those nodes, your fine-grained preferences should be the same as your course-
grained preferences (that is, your preferences over prospects where the final
outcomes do not include information about what could have been chosen in
the decision tree). So, by backward induction, we find that you would accept
the trade at node 3 and therefore that you would accept the trade at node 2.

---

[157] Machina (1989, pp. 1643–4) offers the following example (adapted from Diamond 1967,
pp. 765–6) of a rational fine-grained preference that takes the no-longer-feasible part of the
decision tree into account. A mother with two kids, Abigail and Benjamin, can give one of
them a non-divisible treat. She is indifferent between the sure prospect A, *Abigail getting the
treat*, and the sure prospect B, *Benjamin getting the treat*. Yet she prefers a fifty-fifty gamble
between A and B to each of A and B. So she flips a coin – heads for A; tails for B – and the
coin lands heads. Now, after the coin flip, she prefers *A-given-heads* to *B-given-heads*, which
seems rational. So it's rational to take the no-longer-feasible part of one's decision tree into
account. But this difference in the mother's preference, before and after the coin flip, is more
plausibly due to a preferred difference in the final outcomes of these prospects. Clearly, the
mother prefers distributing the treat fairly, otherwise she wouldn't prefer flipping the coin to
just giving the treat to one of the kids. If her preference is rational, then, given the Principle of
Individuation by Rational Indifference, *Abigail getting the treat unfairly* is not the same final
outcome as *Abigail getting the treat fairly*. Therefore, Machina's example does not show that
it is rational to take the no-longer-feasible part of the decision tree into account. For much the
same response, see Joyce 1999, pp. 53–4.

[158] Nozick (1993, pp. 21–6) objects to the view that sunk costs should be ignored, but see Steele
1996 for a rebuttal.

Taking this into account at node 1, you see that you wouldn't end up with A if you turned down the initial trade. So the choice at node 1 is effectively between A^- and B. Given your fine-grained preference for not being money pumped, we should represent these options as A^--*and-being-money-pumped* and *B-and-not-being-money-pumped*. So, while you prefer A^- to B, you prefer *B-and-not-being-money-pumped* to A^--*and-being-money-pumped*. But, once more, it seems irrational to care, for its own sake, about being money pumped. The problem with being money pumped by paying for A is that you could have kept A for free. But, given your allegedly rational preferences, you predict that you won't keep A. So, since you know at node 1 that you cannot rationally end up with A, you should regard giving up A as a cost that will be sunk. Hence, even in the Upfront Money Pump (and other BI-terminating money pumps), your fine-grained preference for not being money-pumped seems irrational.

Finally, consider the Conservative Approach. By itself, this approach does not violate Decision-Tree Separability, but it does so if it is taken to be a requirement of rationality:[159]

> *The Conservative Approach* When you adopt a plan, you follow that plan as long as you do not prefer not to follow it.

This approach does not help against the Upfront Money Pump, but, as a requirement of rationality, it will block the argument for Minimal Unidimensional Precaution and the arguments that rely on precautionary or dominance backward induction. So it would block the money-pump argument for Completeness which is based on the Precaution Money Pump.

The Conservative Approach does not require that you choose against your preference. So the objection we raised to the Counter-Preferential Approach does not apply here.

But, if you don't prefer following the plan to deviating from it, then it's hard to see what would be irrational about choosing to deviate. Consider, for instance, the choice at node 4 in the Precaution Money Pump, where you have a choice between A and B^-. And suppose that, at the initial node, you adopted the plan to walk away with A. At node 4, you still know that, at the time you adopted the plan, you did not prefer that plan to the one that ends with choosing B^-. The only reason you didn't adopt the plan that ends with choosing B^- was presumably that it is less preferred than the (initially available) plan that ends with choosing B. But this reason no longer applies at node 4 where the

---

[159] Rabinowicz 1995, pp. 594–5. See also Greenberg's (1990, pp. 17–19) similar game-theoretic idea. It is unclear whether Rabinowicz claims that it would be *irrational* to violate the Conservative Approach. What he stresses is that it's rational to expect that one follows the approach at future nodes.

latter plan is unavailable. So it does not seem irrational, at node 4, to deviate from the adopted plan and choose B^-. Of course, choosing B^- at node 4 makes your choice to turn down the trade to B at node 3 seem irrational. But that is not something that affects your rationality at node 4.

8 Against Infinite Money Pumps

Another common objection to money-pump arguments is that they prove too much if we allow infinite decision trees. In cases with infinite series of trade offers, agents with rationally impeccable preferences may be forced to forgo sure monetary benefits. Suppose that you have the following transitive preferences:

(71) ... $A^{+++} > A^{++} > A^+ > A > A^-,$

where A^- is like A except that you have less money, A^+ is like A except that you have more money, A^{++} is like A^+ except that you have even more money, A^{+++} is like A^{++} except that you have still more money, and so on. And suppose that an exploiter will offer you more and more money in an infinite series of trade offers from A to A^+, from A^+ to A^{++}, and so on. After accepting any of these offers, you can walk away with the received money by turning down the next offer. But here's the twist: if you accept all trade offers, you have to give back the money you've received along with an additional payment. Hence, if you accept all trade offers, you end up paying the exploiter for what you could have kept for free. You face the decision problem in Figure 17, the *Infinite Money Pump*.[160]

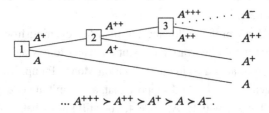

... $A^{+++} > A^{++} > A^+ > A > A^-.$

Figure 17 The Infinite Money Pump

It may seem that, given the preferences in (71), you should accept each trade in the infinite series of trade offers. But, if you do so, you will end up with A^- when you could have kept A for free. And, since the preferences in (71) seem rationally impeccable, this would show that money-pump arguments prove

---

[160] For some variations of this set-up, see Pollock 1983, p. 417, Barrett and Arntzenius 1999, p. 103n3, McGee 1999, pp. 258–60, and Peterson 2016, pp. 167–8.

too much, since these clearly rational preferences would also be vulnerable to exploitation.

Yet this objection can be defused. Note that you would only be rationally required to end up with A^- if you are rationally required to accept each trade. But this is impossible if you use backward induction. Granted, applying backward induction is tricky in decision problems of infinite depth, since it's unclear how the induction could get started without a last node.[161] But, for the following proof by contradiction, we can sidestep this problem, because we merely aim to rule out the possibility that backward induction would prescribe going up at the first node if it were rationally required to go up at each of the later nodes. So suppose, for proof by contradiction, that it's rationally required to accept all trades. Then, using backward induction at node 1, you take into account the prediction that you would accept each of the later trades if you were to accept the first trade. So the choice at node 1 is effectively between A^- (that is, accepting the trade) and A (turning it down). Hence it would be irrational to accept the trade at node 1. This contradicts our initial assumption. Hence it cannot be rationally required to accept all trades.[162]

Could it be rationally *permitted* to accept all trades in the Infinite Money Pump? This is also impossible if agents rely on a preventative version of backward induction:

> According to *preventative backward induction*, it is irrational to choose an option X over an option Y if there is a rationally allowed outcome of X (that is, a prospect of an available plan consisting in choosing X followed by choices that are not irrational) that is less preferred than any rationally allowed outcome of Y.

Suppose, for proof by contradiction, that it's rationally permitted to accept all trades. Then, at node 1, there is a rationally permitted sequence of choices following the choice to accept the first trade such that accepting that trade leads to your ending up with A^-, which is less preferred than the only rationally allowed outcome of turning down the first trade. So, given preventative backward induction, it's not rationally permitted to accept the first trade. This contradicts our initial assumption. Hence it cannot be rationally permitted to accept all trades.

Note that this argument does not show that it must be rationally required to turn down the first trade or, more generally, to terminate any decision

---

[161] Sobel 2001, p. 256.

[162] Arntzenius, Elga, and Hawthorne (2004, p. 267) claim that, if you cannot bind yourself irrevocably to a certain plan and you cannot influence your future choices, then you are required to accept all trade offers and end up with A^-. But my argument does not assume that you can influence your future choices nor that you can bind yourself irrevocably to a plan.

problem at the first node if the problem has this general preference pattern over outcomes. If you did so, you would be vulnerable to an upfront variation of the Infinite Money Pump in case you had the following, clearly rational, preferences:

(72) ... $A^{+++} > A^{++} > A^{+} > A > A^{-} > A^{--}$.

Consider the variation in Figure 18, the *Upfront Infinite Money Pump*, where we have added an initial opportunity to pay the exploiter to go away and this payment will be slightly lower than the amount you pay in case you accept all the later trade offers.

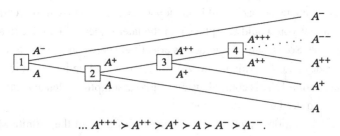

... $A^{+++} > A^{++} > A^{+} > A > A^{-} > A^{--}$.

Figure 18 The Upfront Infinite Money Pump

Here, just like in the Infinite Money Pump, each non-terminal choice leads to a choice node with a more preferred prospect for the next terminal choice. But, if you choose the terminal option at node 1 in the Upfront Infinite Money Pump, you end up with A^{-} when you could have kept A for free.

So what should you do in these infinite cases? In the following, I will merely sketch a potential answer. The idea is to think of potential rationales for choosing in the decision problem. Discard any rationale you wouldn't find compelling at the future choice nodes you could face if you were to choose based on that rationale. And then choose based on a rationale (among the remaining rationales) such that the prospect of choosing based on that rationale is at least as preferred as the prospect of choosing based on any of the other remaining rationales.

There seems to be a compelling rationale for walking away with A. In the Upfront Infinite Money Pump, you turn down the trade from A to A^{-}, since you do not want to pay for what you could keep for free. The trouble with any rationale for accepting any further trades (to A^{+}, A^{++}, and so on) is that they either support accepting all trades, which leads to exploitation, or they will support turning down any further trades once you have A with a certain number of pluses. The trouble with the latter is that, since there is nothing special about A with that specific number of pluses, any rationale that supports trading until you have *A* with that many pluses would be arbitrary. So you wouldn't find this rationale compelling once it supports walking way.[163]

Note that, in both the Infinite Money Pump and the Upfront Infinite Money Pump, you can't avoid forgoing a sure benefit. So, if money-pump arguments take forgoing sure benefits to be a sign of irrationality, then they would prove too much. This is so, since the clearly rational preferences in (71) and (72) force the agent to forgo a sure benefit in these decision problems.[164] But money-pump arguments do not prove too much if they merely take exploitability to be a sign of irrationality.

---

[163] Why not plumb for the largest salient finite prospect (say *A* with googolplex pluses), then make the choices necessary to get that prospect, and then stop there? The trouble is that the plan to stop after a googolplex steps was chosen arbitrarily at the start, so why would you care about that plan when you reach the googolplexth node? You would prefer to go on for another googolplex steps. And so on. This problem also rules out using randomization at the start to determine how many pluses you should walk away with. A better idea, if you are able to randomize, is to perform the mixed strategy of turning down the offers with the lowest salient non-zero probability. As long as there is some chance that you turn down the offers, you are guaranteed to eventually walk away and thereby avoid exploitation. But this idea only works if you would in fact turn down the next trade once randomness favours turning down the next trade. The danger is that you may be tempted to roll the dice again.

[164] This may also happen in non-sequential situations if there are infinitely many alternatives available at the same time – each less preferred than some other alternative. (See Savage 1954, p. 18, Nozick 1963, p. 89, and Gustafsson 2013, p. 464.) In such situations, you may be forced to forgo sure benefits whatever you do, but you can still avoid exploitation.

Appendices

A Notation

$X \gtrsim Y =_{\mathrm{df}} X$ is at least as preferred to Y

$X > Y =_{\mathrm{df}} X \gtrsim Y$ and it is not the case that $Y \gtrsim X$.

$X \sim Y =_{\mathrm{df}} X \gtrsim Y$ and $Y \gtrsim X$.

$X \parallel Y =_{\mathrm{df}}$ it is neither the case that $X \gtrsim Y$ nor the case that $Y \gtrsim X$.

$XpY =_{\mathrm{df}}$ a prospect consisting in a lottery between X and Y such that X occurs with probability p and Y occurs with probability $1 - p$.

B Principles

Acyclicity If $X_1 > X_2 > \ldots > X_n$, then it is not the case that $X_n > X_1$.

Completeness $X \gtrsim Y$ or $Y \gtrsim X$.

Continuity If $X > Y > Z$, then there are probabilities p and q such that (i) $0 < p < 1$, (ii) $0 < q < 1$, and (iii) $XpZ > Y > XqZ$.

Decision-Tree Separability The rational status of the options at a choice node does not depend on other parts of the decision tree than those that can be reached from that node.

Independence (the biconditional weak-preference version) For all probabilities p such that $0 < p < 1$, it holds that $X \gtrsim Y$ if and only if $XpZ \gtrsim YpZ$.

Independence (the strong strict-preference version) For all probabilities p such that $0 < p < 1$, it holds that, if $X > Y$, then $XpZ > YpZ$.

Independence (the weak strict-preference version) For all probabilities p such that $0 < p < 1$, it holds that, if $X > Y$, then it is not the case that $YpZ > XpZ$.

Independence for Constant Outcomes (the weak strict-preference version)
For all probabilities p such that $0 < p < 1$, it holds that, if $XpU > YpU$, then it is not the case that $YpV > XpV$.

The Irrationality of Single Sourings   If (i) X^- is a souring of X, (ii) $X > X^-$, (iii) node n is a choice between node n^* and X, (iv) node n^* is a choice between X^- and Y, and (v) one knows at node n what decision problem one faces, then the sequence of choices consisting in choosing node n^* at node n and X^- at node n^* violates a requirement of rationality.

The Maximization Rule   It is rationally permitted to choose a prospect X if and only if there is no feasible prospect Y such that $Y > X$.

Minimal Unidimensional Precaution   If (i) X^- is a souring of X, (ii) $X > X^-$, (iii) it is not the case that $Y > X^-$, (iv) node n is a choice between node n^* and X, (v) node n^* is a choice between X^- and Y, and (vi) one knows at node n what decision problem one faces, then one chooses X at node n.

One-Step Acyclicity   It is not the case that $X > X$.

The Principle of Future-Choice Independence   The rational status of an option at a choice node and the rational status of the agent's credences and preferences at that node do not depend on what would in fact be chosen at later choice nodes.

The Principle of Individuation by Rational Indifference   Final outcomes x and y should be treated as the same if and only if it is rationally required to be indifferent between the sure prospects of x and y.

The Principle of Limit Unexploitability   If (i) ϵ is a fixed positive amount, (ii) X^- is a souring of X such that X^- is certainly ϵ units inferior to X in a dimension one cares about, (iii) $X > X^-$, (iv), at node n, P and P^- are two available plans such that P is the only available plan that amounts to walking away from all offers by an exploiter and the prospect of following P is X and the prospect of following P^- is arbitrarily likely to be X^-, and (v) one knows what decision problem one faces at n, then one does not follow P^- from n.

The Principle of Preferential Invulnerability   If there is a possible situation where having a certain combination of preferences forces one to violate a requirement of rationality, then there is a requirement of rationality that rules out that combination of preferences in all possible situations.

The Principle of Rational Decomposition   If an agent, whose credences and preferences are not rationally prohibited, makes a sequence of choices which violates a requirement of rationality, then some of those choices are rationally prohibited.

The Principle of Unexploitability   If (i) X^- is a souring of X, (ii) $X > X^-$, (iii), at node n, it holds that P and P^- are two available plans such that P is the only available plan that amounts to walking away from all offers by an exploiter and the prospect of following P is X and the prospect of following P^- is X^-, and (iv) one knows what decision problem one faces at n, then one does not follow P^- from n.

The Souring Principle   If X^- is a souring of X, then $X > X^-$.

Strong Acyclicity   If $X_1 \gtrsim X_2 \gtrsim \ldots \gtrsim X_n$, then it is not the case that $X_n > X_1$.

Strong Insensitivity to Souring   If $X \parallel Y$, then there is a prospect X^- such that (i) X^- is a souring of X and (ii) $X > X^- \parallel Y$.

The Strong Principle of Eventwise Dominance   If there is a set of events such that (i) the set is a partition of states of nature, (ii), given each event E in the set, the outcome of gamble G given E is at least as preferred as the outcome of gamble G^* given E, and (iii), in some positive-probability event E^* in the set, the outcome of G given E^* is preferred to the outcome of G^* given E^*, then $G > G^*$.

The Strong Principle of Unidimensional Stochastic Dominance   If (i) X^- is a souring of X, (ii) $X > X^-$, and (iii) $0 < p \leq 1$, then $XpY > X^-pY$.

Symmetry of Souring Sensitivity   If (i) X^- is a souring of X and (ii) $X > X^- \parallel Y \parallel X$, then there is a prospect Y^- such that (i) Y^- is a souring of Y and (ii) $Y > Y^- \parallel X$.

Three-Step Acyclicity   If $X > Y > Z$, then it is not the case that $Z > X$.

Transitivity   If $X \gtrsim Y \gtrsim Z$, then $X \gtrsim Z$.

Two-Step Acyclicity   If $X > Y$, then it is not the case that $Y > X$.

The Uncovered-Choice Rule   It is rationally permitted to choose a prospect X if and only if there is no feasible prospect Y such that $Y > X$ and, for all feasible prospects Z, it holds that $Y > Z$ if $X > Z$.

Unidimensional Continuity of Dispreference   If $X > Y$, then there is a prospect Y^+ such that (i) Y^+ is a sweetening of Y and (ii) $X > Y^+ > Y$.

Unidimensional Continuity of Preference   If $X > Y$, then there is a prospect X^- such that (i) X^- is a souring of X and (ii) $X > X^- > Y$.

Unidimensional IP-Transitivity   If (i) Y^- is a souring of Y and (ii) $X \sim Y >$ Y^-, then $X > Y^-$.

Unidimensional PI-Acyclicity   If (i) X^+ is a sweetening of X and (ii) $X^+ >$ $X \sim Y$, then it is not the case that $Y > X^+$.

Weak Insensitivity to Souring   If $X \parallel Y$, then

- there is a prospect X^- such that (i) X^- is a souring of X and (ii) $X >$ $X^- \parallel Y$ or
- there is a prospect Y^- such that (i) Y^- is a souring of Y and (ii) $Y >$ $Y^- \parallel X$.

The Weak Principle of Equiprobable Unidimensional Dominance   If there are sets of events $\{E_1, E_2, \ldots\}$ and $\{E_1^*, E_2^*, \ldots\}$ such that these sets are partitions of states of nature and, for all $i = 1, 2, \ldots$, it holds that (a) E_i has the same probability as E_i^*, (b) the outcome of gamble G^* given E_i^* is a souring of the outcome of gamble G given E_i, and (c) the outcome of G given E_i is preferred to the outcome of G^* given E_i^*, then $G > G^*$.

References

Aaronson, Scott (2016) 'The Ghost in the Quantum Turing Machine', in S. Barry Cooper and Andrew Hodges, eds., *The Once and Future Turing: Computing the World*, pp. 193–296, Cambridge: Cambridge University Press.

Ahmed, Arif (2014) *Evidence, Decision and Causality*, Cambridge: Cambridge University Press.

(2017) 'Exploiting Cyclic Preference', *Mind* 126 (504): 975–1022.

Al-Najjar, Nabil I. and Jonathan Weinstein (2009) 'The Ambiguity Aversion Literature: A Critical Assessment', *Economics and Philosophy* 25 (3): 249–84.

Alchian, Armen A. (1953) 'The Meaning of Utility Measurement', *The American Economic Review* 43 (1): 26–50.

Allais, Maurice (1953) 'Le comportement de l'homme rationnel devant le risque: Critique des postulats et axiomes de l'ecole Americaine', *Econometrica* 21 (4): 503–46.

(1979) 'The Foundations of a Positive Theory of Choice involving Risk and a Criticism of the Postulates and Axioms of the American School', in M. Allais and O. Hagen, eds., *Expected Utility Hypothesis and the Allais Paradox: Contemporary Discussions of Decisions under Uncertainty with Allais' Rejoinder*, pp. 27–145, Dordrecht: Reidel.

Anand, Paul (1987) 'Are the Preference Axioms Really Rational?', *Theory and Decision* 23 (2): 189–214.

(1990) 'Interpreting Axiomatic (Decision) Theory', *Annals of Operations Research* 23 (1): 91–101.

(1993a) *Foundations of Rational Choice under Risk*, Oxford: Clarendon Press.

(1993b) 'The Philosophy of Intransitive Preference', *The Economic Journal* 103 (417): 337–46.

Arkes, Hal R., Gerd Gigerenzer, and Ralph Hertwig (2016) 'How Bad Is Incoherence?', *Decision* 3 (1): 20–39.

Armstrong, W. E. (1939) 'The Determinateness of the Utility Function', *The Economic Journal*, pp. 453–67.

Arnauld, Antoine and Pierre Nicole (1662) *La logique ou l'art de penser: Contenant, outre les regles communes, plusieurs observations nouvelles, propres à former le jugement*, Paris: Jean Guignart, Charles Savreux, & Jean de Lavnay.

(1996) *Logic or the Art of Thinking: Containing, besides Common Rules, Several New Observations Appropriate for Forming Judgment*, ed. Jill Vance Buroker, Cambridge: Cambridge University Press.

Arntzenius, Frank, Adam Elga, and John Hawthorne (2004) 'Bayesianism, Infinite Decisions, and Binding', *Mind* 113 (450): 251–83.

Arrow, Kenneth J. (1951) *Social Choice and Individual Values*, New York: Wiley.

(1965) *Aspects of the Theory of Risk-Bearing*, Helsinki: Yrjö Jahnssonin Säätiö.

Aumann, Robert J. (1962) 'Utility Theory without the Completeness Axiom', *Econometrica* 30 (3): 445–62.

(1998) 'On the Centipede Game', *Games and Economic Behavior* 23 (1): 97–105.

Bader, Ralf M. (2019) 'Agent-Relative Prerogatives and Suboptimal Beneficence', in Mark Timmons, ed., *Oxford Studies in Normative Ethics Volume 9*, pp. 223–50, Oxford: Oxford University Press.

Barrett, Jeffrey A. and Frank Arntzenius (1999) 'An Infinite Decision Puzzle', *Theory and Decision* 46 (1): 101–3.

Bernoulli, Daniel (1738) 'Specimen Theoriae Novae de Mensura Sortis', *Commentarii Academiae Scientiarum Imperialis Petropolitana* 5 (1): 175–92.

(1954) 'Exposition of a New Theory on the Measurement of Risk', *Econometrica* 22 (1): 23–36.

Binmore, Ken (1987) 'Modeling Rational Players: Part I', *Economics and Philosophy* 3 (2): 179–214.

Black, Duncan (1948) 'On the Rationale of Group Decision-Making', *Journal of Political Economy* 56 (1): 23–4.

Blackwell, David and M. A. Girshick (1954) *Theory of Games and Statistical Decisions*, New York: Wiley.

Bossert, Walter, Yves Sprumont, and Kotaro Suzumura (2005) 'Consistent Rationalizability', *Economica* 72 (286): 185–200.

Bostrom, Nick (2011) 'Infinite Ethics', *Analysis and Metaphysics* 10 (1): 9–59.

Bratman, Michael E. (1992) 'Planning and the Stability of Intention', *Minds and Machines* 2 (1): 1–16.

(1998) 'Following Through with One's Plans: Reply to David Gauthier', in Peter A. Danielson, ed., *Modeling Rationality, Morality, and Evolution*, pp. 55–66, New York: Oxford University Press.

Briggs, Rachael (2010) 'Putting a Value on Beauty', in Tamar Szabó Gendler and John Hawthorne, eds., *Oxford Studies in Epistemology Volume 3*, pp. 3–34, Oxford: Oxford University Press.

Broome, John (1990) 'Should a Rational Agent Maximize Expected Utility?', in Karen Schweers Cook and Margaret Levi, eds., *The Limits of Rationality*, pp. 132–45, Chicago: University of Chicago Press.

(1991) *Weighing Goods: Equality, Uncertainty and Time*, Oxford: Blackwell.

(1992) 'Review of Edward F. McClennen, *Rationality and Dynamic Choice: Foundational Explorations*', *Ethics* 102 (3): 666–668.

(1993) 'Can a Humean Be Moderate?', in R. G. Frey and Christopher W. Morris, eds., *Value, Welfare, and Morality*, pp. 51–73, Cambridge: Cambridge University Press.

(1999) *Ethics Out of Economics*, Cambridge: Cambridge University Press.

(2000) 'Incommensurable Values', in Roger Crisp and Brad Hooker, eds., *Well-Being and Morality: Essays in Honour of James Griffin*, pp. 21–38, Oxford: Clarendon Press.

(2004) *Weighing Lives*, Oxford: Oxford University Press.

Broome, John and Wlodek Rabinowicz (1999) 'Backwards Induction in the Centipede Game', *Analysis* 59 (4): 237–42.

Buchak, Lara (2013) *Risk and Rationality*, Oxford: Oxford University Press.

Burros, Raymond H. (1974) 'Axiomatic Analysis of Non-Transitivity of Preference and of Indifference', *Theory and Decision* 5 (2): 185–204.

Cantwell, John (2002) 'The Pragmatic Stance: Whither Dutch Books and Money Pumps?', *Croatian Journal of Philosophy* 2 (6): 319.

(2003) 'On the Foundations of Pragmatic Arguments', *The Journal of Philosophy* 100 (8): 383–402.

Cayley, Arthur (1875) 'Mathematics Problem 4528', *The Educational Times* 27 (166): 237.

Chang, Ruth (1997) 'Introduction', in Ruth Chang, ed., *Incommensurability, Incomparability, and Practical Reason*, pp. 1–34, Cambridge, MA: Harvard University Press.

Chisholm, Roderick M. and Ernest Sosa (1966) 'On the Logic of "Intrinsically Better" ', *American Philosophical Quarterly* 3 (3): 244–9.

Christensen, David (1996) 'Dutch-Book Arguments Depragmatized: Epistemic Consistency for Partial Believers', *The Journal of Philosophy* 93 (9): 450–79.

Condorcet, Marquis de (1785) *Essai sur l'application de l'analyse à la probabilité des décisions rendues à la pluralité des voix*, Paris: Imprimerie royale.

(1976) 'Essay on the Application of Mathematics to the Theory of Decision-Making', in Keith Michael Baker, ed., *Selected Writings*, pp. 33–70, Indianapolis: Bobbs-Merrill.

(1994) 'An Essay on the Application of Probability to Plurality Decision-Making', in Iain McLean and Fiona Hewitt, eds., *Condorcet: Foundations of Social Choice and Political Theory*, pp. 120–38, Aldershot: Edward Elgar.

Cubitt, Robin P. and Robert Sugden (2001) 'On Money Pumps', *Games and Economic Behavior* 37 (1): 121–60.

Davidson, Donald, J. C. C. McKinsey, and Patrick Suppes (1955) 'Outlines of a Formal Theory of Value, I', *Philosophy of Science* 22 (2): 140–60.

De Morgan, Augustus (1851) 'On the Symbols of Logic, the Theory of the Syllogism, and in Particular of the Copula, and the Application of the Theory of Probabilities to Some Questions of Evidence', *Transactions of the Cambridge Philosophical Society* 9 (1): 79–127.

Debreu, Gerard (1954) 'Representation of a Preference Ordering by a Numerical Function', in R. M. Thrall, C. H. Coombs, and R. L. Davis, eds., *Decision Processes*, pp. 159–65, New York: Wiley.

(1959) *Theory of Value: An Axiomatic Analysis of Economic Equilibrium*, New York: Wiley.

Diamond, Peter A. (1967) 'Cardinal Welfare, Individualistic Ethics, and Interpersonal Comparisons of Utility: Comment', *The Journal of Political Economy* 75 (5): 756–66.

Dodgson, Charles Lutwidge (1876) *A Method of Taking Votes on More Than Two Issues*, Oxford: Clarendon Press.

Dow, Gregory K. (1984) 'Myopia, Amnesia, and Consistent Intertemporal Choice', *Mathematical Social Sciences* 8 (2): 95–109.

Dummett, Michael (1984) *Voting Procedures*, Oxford: Clarendon Press.

Edwards, Ward, Harold Lindman, and Lawrence D. Phillips (1965) 'Emerging Technologies for Making Decisions', in Theodore M. Newcomb, ed., *New Directions in Psychology II*, pp. 259–325, New York: Holt, Rinehart and Winston.

Elga, Adam (2010) 'Subjective Probabilities Should Be Sharp', *Philosophers' Imprint* 10 (5): 1–11.

Ellsberg, Daniel (1961) 'Risk, Ambiguity, and the Savage Axioms', *The Quarterly Journal of Economics* 75 (4): 643–69.

Etchart, Nathalie (2002) 'Adequate Moods for Non-EU Decision Making in a Sequential Framework: A Synthetic Discussion', *Theory and Decision* 52 (1): 1–28.

Feynman, Richard P., Robert B. Leighton, and Matthew Sands (1963) *The Feynman Lectures on Physics, Volume I: Mainly Mechanics, Radiation, and Heat*, Reading, MA: Addison-Wesley.

Fishburn, Peter C. (1970) *Utility Theory for Decision Making*, New York: Wiley.

(1977) 'Condorcet Social Choice Functions', *SIAM Journal on Applied Mathematics* 33 (3): 469–89.

(1991) 'Nontransitive Preferences in Decision Theory', *Journal of Risk and Uncertainty* 4 (2): 113–34.

Fishburn, Peter C. and Peter P. Wakker (1995) 'The Invention of the Independence Condition for Preferences', *Management Science* 41 (7): 1130–44.

Gardner, Martin (1970) 'The Paradox of the Nontransitive Dice and the Elusive Principle of Indifference', *Scientific American* 223 (6): 110–15.

Gauthier, David (1996) 'Commitment and Choice: An Essay on the Rationality of Plans', in Francesco Farina, Frank Hahn, and Stefano Vannucci, eds., *Ethics, Rationality, and Economic Behaviour*, pp. 217–43, Oxford: Clarendon Press.

Ginet, Carl (1990) *On Action*, Cambridge: Cambridge University Press.

Goldman, Holly S. (1978) 'Doing the Best One Can', in Alvin I. Goldman and Jaegwon Kim, eds., *Values and Morals*, pp. 185–214, Dordrecht: Reidel.

Greenberg, Joseph (1990) *The Theory of Social Situations: An Alternative Game-Theoretic Approach*, Cambridge: Cambridge University Press.

Gustafsson, Johan E. (2010) 'A Money-Pump for Acyclic Intransitive Preferences', *Dialectica* 64 (2): 251–7.

(2013) 'The Irrelevance of the Diachronic Money-Pump Argument for Acyclicity', *The Journal of Philosophy* 110 (8): 460–4.

(2016) 'Money Pumps, Incompleteness, and Indeterminacy', *Philosophy and Phenomenological Research* 92 (1): 60–72.

(2018) 'Bentham's Binary Form of Maximizing Utilitarianism', *British Journal for the History of Philosophy* 26 (1): 87–109.

(2020) 'Permissibility Is the Only Feasible Deontic Primitive', *Philosophical Perspectives* 34 (1): 117–33.

(2021) 'The Sequential Dominance Argument for the Independence Axiom of Expected Utility Theory', *Philosophy and Phenomenological Research* 103 (1): 21–39.

Gustafsson, Johan E. and Nicolas Espinoza (2010) 'Conflicting Reasons in the Small-Improvement Argument', *The Philosophical Quarterly* 60 (241): 754–63.

Gustafsson, Johan E. and Wlodek Rabinowicz (2020) 'A Simpler, More Compelling Money Pump with Foresight', *The Journal of Philosophy* 117 (10): 578–89.

Halldén, Sören (1957) *On the Logic of 'Better'*, Lund: Gleerup.

Halstead, John (2015) 'The Impotence of the Value Pump', *Utilitas* 27 (2): 195–216.

Hammond, Peter J. (1976) 'Changing Tastes and Coherent Dynamic Choice', *The Review of Economic Studies* 43 (1): 159–73.

(1988a) 'Consequentialism and the Independence Axiom', in Bertrand R. Munier, ed., *Risk, Decision and Rationality*, pp. 503–16, Dordrecht: Reidel.

(1988b) 'Orderly Decision Theory: A Comment on Professor Seidenfeld', *Economics and Philosophy* 4 (2): 292–7.

(1998) 'Objective Expected Utility: A Consequentialist Perspective', in Salvador Barberà, Peter J. Hammond, and Christian Seidl, eds., *Handbook of Utility Theory Volume 1: Principles*, pp. 143–211, Dordrecht: Kluwer.

Hansson, Sven Ove (1993) 'Money-Pumps, Self-Torturers and the Demons of Real Life', *Australasian Journal of Philosophy* 71 (4): 476–85.

(2002) 'Preference Logic', in Dov M. Gabbay and F. Guenthner, eds., *Handbook of Philosophical Logic*, vol. 4, pp. 319–93, Dordrecht: Kluwer, 2nd ed.

Hedden, Brian (2015a) 'Options and Diachronic Tragedy', *Philosophy and Phenomenological Research* 90 (2): 423–51.

(2015b) *Reasons without Persons: Rationality, Identity, and Time*, Oxford: Oxford University Press.

Herstein, I. N. and John Milnor (1953) 'An Axiomatic Approach to Measurable Utility', *Econometrica* 21 (2): 291–97.

Homer (1995) *Odyssey: Books 1–12*, edited by A. T. Murray and George E. Dimock, Loeb Classical Library, Cambridge, MA: Harvard University Press, 2nd ed.

Houthakker, H. S. (1950) 'Revealed Preference and the Utility Function', *Economica* 17 (66): 159–74.

Jensen, Niels Erik (1967) 'An Introduction to Bernoullian Utility Theory: I. Utility Functions', *Swedish Journal of Economics* 69 (3): 163–83.

Joyce, James M. (1999) *The Foundations of Causal Decision Theory*, Cambridge: Cambridge University Press.

Kahneman, Daniel and Amos Tversky (1979) 'Prospect Theory: An Analysis of Decision under Risk', *Econometrica* 47 (2): 263–91.

Kohlberg, Elon and Jean-Francois Mertens (1986) 'On the Strategic Stability of Equilibria', *Econometrica* 54 (5): 1003–37.

Kowalczyk, Kacper (2020) 'Pumping Discontinuity', unpublished manuscript.

Kripke, Saul (1972) 'Naming and Necessity', in Donald Davidson and Gilbert Harman, eds., *Semantics of Natural Language*, pp. 253–355, Dordrecht: Reidel.

Lehman, R. Sherman (1955) 'On Confirmation and Rational Betting', *Journal of Symbolic Logic* 20 (3): 251–62.

Loomes, Graham and Robert Sugden (1987) 'Some Implications of a More General Form of Regret Theory', *Journal of Economic Theory* 41 (2): 270–87.

Luce, R. Duncan (1956) 'Semiorders and a Theory of Utility Discrimination', *Econometrica* 24 (2): 178–91.

Luce, R. Duncan and Howard Raiffa (1957) *Games and Decisions: Introduction and Critical Survey*, New York: Wiley.

Machina, Mark J. (1989) 'Dynamic Consistency and Non-Expected Utility Models of Choice under Uncertainty', *Journal of Economic Literature* 27 (4): 1622–68.

Marschak, Jakob (1950) 'Rational Behavior, Uncertain Prospects, and Measurable Utility', *Econometrica* 18 (2): 111–41.

May, Kenneth O. (1954) 'Intransitivity, Utility, and the Aggregation of Preference Patterns', *Econometrica* 22 (1): 1–13.

McClennen, Edward F. (1985) 'Prisoner's Dilemma and Resolute Choice', in Richmond Campbell and Lanning Sowden, eds., *Paradoxes of Rationality and Cooperation: Prisoner's Dilemma and Newcomb's Problem*, pp. 94–104, Vancouver: University of British Columbia Press.

(1988) 'Dynamic Choice and Rationality', in Bertrand R. Munier, ed., *Risk, Decision and Rationality*, pp. 517–36, Dordrecht: Reidel.

(1990) *Rationality and Dynamic Choice: Foundational Explorations*, Cambridge: Cambridge University Press.

(1997) 'Pragmatic Rationality and Rules', *Philosophy & Public Affairs* 26 (3): 210–58.

(2009) 'Exploitable Preference Changes', in Till Grüne-Yanoff and Sven Ove Hansson, eds., *Preference Change: Approaches from Philosophy, Economics and Psychology*, pp. 123–37, Berlin: Springer.

McClennen, Edward F. and Scott Shapiro (1998) 'Rule-Guided Behaviour', in Peter Newman, ed., *The New Palgrave Dictionary of Economics and the Law*, vol. 3, pp. 363–9, London: Macmillan.

McGee, Vann (1999) 'An Airtight Dutch Book', *Analysis* 59 (4): 257–65.

McMahan, Jefferson (1981) 'Problems of Population Theory', *Ethics* 92 (1): 96–127.

Miller, Nicholas R. (1980) 'A New Solution Set for Tournaments and Majority Voting: Further Graph-Theoretical Approaches to the Theory of Voting', *American Journal of Political Science* 24 (1): 68–96.

Milnor, John (1954) 'Games against Nature', in Robert M. Thrall, Clyde H. Coombs, and Robert L. Davis, eds., *Decision Processes*, pp. 49–59, New York: Wiley.

Mongin, Philippe (1999) 'The Allais Paradox: What It Became, What It Really Was, What It Now Suggests to Us', *Economics and Philosophy* 35 (3): 423–59.

Morlat, Georges (1953) 'Intervention de M. Morlat', in *Économétrie*, pp. 156–7, Paris: Centre national de la recherche scientifique.

Moulin, H. (1983) *The Strategy of Social Choice*, Amsterdam: North-Holland.

Nagel, T. (1976) 'Moral Luck', *Proceedings of the Aristotelian Society, Supplementary Volumes* 50 (1):137–51.

Nash, John F. Jr. (1950) 'The Bargaining Problem', *Econometrica* 18 (2): 155–62.

Ng, Yew-Kwang (1975) 'Bentham or Bergson? Finite Sensibility, Utility Functions and Social Welfare Functions', *The Review of Economic Studies* 42 (4): 545–69.

(1977) 'Sub-Semiorder: A Model of Multidimensional Choice with Preference Intransitivity', *Journal of Mathematical Psychology* 16 (1): 51–9.

Nozick, Robert (1963) *The Normative Theory of Individual Choice*, Ph.D. thesis, Princeton University.

(1993) *The Nature of Rationality*, Princeton: Princeton University Press.

Parfit, Derek (1982) 'Future Generations: Further Problems', *Philosophy & Public Affairs* 11 (2): 113–72.

(2011) *On What Matters*, vol. 1, Oxford: Oxford University Press.

Pascal, Blaise (1670) *Pensées de M. Pascal sur la religion et sur quelques autres sujets, qui ont été trouvées après sa mort parmi ses papiers*, Paris: Guillaume Desprez.

(2004) *Pensées*, ed. Roger Ariew, Indianapolis: Hackett.

Peterson, Martin (2007) 'Parity, Clumpiness and Rational Choice', *Utilitas* 19 (4): 505–13.

(2016) 'Do Pragmatic Arguments Show Too Much?', *European Journal for Philosophy of Science* 6 (2): 165–72.

Pettigrew, Richard (2020) *Dutch Book Arguments*, Cambridge: Cambridge University Press.

Pollak, R. A. (1968) 'Consistent Planning', *The Review of Economic Studies* 35 (2): 201–8.

Pollock, John L. (1983) 'How Do You Maximize Expectation Value?', *Noûs* 17 (3): 409–21.

Pratt, John W., Howard Raiffa, and Robert Schlaifer (1965) *Introduction to Statistical Decision Theory*, New York: McGraw-Hill, preliminary ed.

Priest, Graham (2005) *Towards Non-Being: The Logic and Metaphysics of Intentionality*, Oxford: Clarendon Press.

Quinn, Warren S. (1990) 'The Puzzle of the Self-Torturer', *Philosophical Studies* 59 (1): 79–90.

Rabinowicz, Wlodek (1995) 'To Have One's Cake and Eat It, Too: Sequential Choice and Expected-Utility Violations', *The Journal of Philosophy* 92 (11): 586–620.

    (1998) 'Grappling with the Centipede: Defence of Backward Induction for BI-Terminating Games', *Economics and Philosophy* 14 (1): 95–126.

    (2000a) 'Money Pump with Foresight', in Michael J. Almeida, ed., *Imperceptible Harms and Benefits*, pp. 123–54, Dordrecht: Kluwer.

    (2000b) 'Preference Stability and Substitution of Indifferents: A Rejoinder to Seidenfeld', *Theory and Decision* 48 (4): 311–18.

    (2008) 'Pragmatic Arguments for Rationality Constraints', in Maria Carla Galavotti, Roberto Scazzieri, and Patrick Suppes, eds., *Reasoning, Rationality, and Probability*, pp. 139–63, Stanford, CA: CSLI.

    (2012) 'Value Relations Revisited', *Economics and Philosophy* 28 (2): 133–64.

Raiffa, Howard (1961) 'Risk, Ambiguity, and the Savage Axioms: Comment', *The Quarterly Journal of Economics* 75 (4): 690–94.

    (1968) *Decision Analysis: Introductory Lectures on Choices under Uncertainty*, Reading, MA: Addison-Wesley.

Raiffa, Howard and Robert Schlaifer (1961) *Applied Statistical Decision Theory*, Boston: Harvard Business School.

Ramsey, Frank Plumpton (1931) 'Truth and Probability', in R. B. Braithwaite, ed., *The Foundations of Mathematics and Other Logical Essays*, pp. 156–98, London: Routledge and Kegan Paul.

Raz, Joseph (1985–6) 'Value Incommensurability: Some Preliminaries', *Proceedings of the Aristotelian Society* 86 (1): 117–34.

    (1986) *The Morality of Freedom*, Oxford: Clarendon Press.

Regan, Donald H. (2000) 'Perceiving Imperceptible Harms (with Other Thoughts on Transitivity, Cumulative Effects, and Consequentialism)', in Michael J. Almeida, ed., *Imperceptible Harms and Benefits*, pp. 49–73, Dordrecht: Kluwer.

Restle, Frank (1961) *Psychology of Judgment and Choice: A Theoretical Essay*, New York: Wiley.

Roberts, Fred Stephen (1979) *Measurement Theory with Applications to Decisionmaking, Utility, and the Social Sciences*, Reading, MA: Addison-Wesley.

Rubin, Herman (1949) 'The Existence of Measurable Utility and Psychological Probability', Cowles Commission Discussion Paper: Statistics: No. 331.

Russell, Bertrand (1903) *The Principles of Mathematics Vol I.*, Cambridge: Cambridge University Press.

Samuelson, Paul Anthony (1947) *Foundations of Economic Analysis*, Cambridge, MA: Harvard University Press.

Sartre, Jean-Paul (1946) *L'existentialism est un humanisme*, Paris: Les Èditions Nagel.

(2007) *Existentialism Is a Humanism*, ed. John Kulka, New Haven, CT: Yale University Press.

Savage, Leonard J. (1951) 'The Theory of Statistical Decision', *Journal of the American Statistical Association* 46 (253): 55–67.

(1954) *The Foundations of Statistics*, New York: Wiley.

Schick, Frederic (1986) 'Dutch Bookies and Money Pumps', *The Journal of Philosophy* 83 (2): 112–19.

Schwartz, Thomas (1970) 'On the Possibility of Rational Policy Evaluation', *Theory and Decision* 1 (1): 89–106.

(1986) *The Logic of Collective Choice*, New York: Columbia University Press.

(1990) 'Cyclic Tournaments and Cooperative Majority Voting: A Solution', *Social Choice and Welfare* 7 (1): 19–29.

Seidenfeld, Teddy (1988) 'Decision Theory without "Independence" or without "Ordering": What Is the Difference?', *Economics and Philosophy* 4 (2): 267–90.

(2000) 'Substitution of Indifferent Options at Choice Nodes and Admissibility: A Reply to Rabinowicz', *Theory and Decision* 48 (4): 305–10.

Selten, Reinhard (1978) 'The Chain Store Paradox', *Theory and Decision* 9 (2): 127–59.

Sen, Amartya K. (1977) 'Social Choice Theory: A Re-Examination', *Econometrica* 45 (1): 53–89.

Sobel, Jordan Howard (1976) 'Utilitarianism and Past and Future Mistakes', *Noûs* 10 (2): 195–219.

(1993) 'Backward-Induction Arguments: A Paradox Regained', *Philosophy of Science* 60 (1): 114–33.

(2001) 'Money Pumps', *Philosophy of Science* 68 (2): 242–57.

Sonnenschein, Hugo (1965) 'The Relationship between Transitive Preference and the Structure of the Choice Space', *Econometrica* 33 (3): 624–34.

Steele, David Ramsay (1996) 'Nozick on Sunk Costs', *Ethics* 106 (3): 605–20.

Steele, Katie (2007) *Precautionary Decision-Making: An Examination of Bayesian Decision Norms in the Dynamic Choice Context*, Ph.D. thesis, University of Sydney.

(2010) 'What Are the Minimal Requirements of Rational Choice? Arguments from the Sequential-Decision Setting', *Theory and Decision* 68 (4): 463–87.

(2018) 'Dynamic Decision Theory', in Sven Ove Hansson and Vincent F. Hendricks, eds., *Introduction to Formal Philosophy*, pp. 657–67, Berlin: Springer.

Strotz, R. H. (1955–6) 'Myopia and Inconsistency in Dynamic Utility Maximization', *The Review of Economic Studies* 23 (3): 165–80.

Suzumura, Kotaro (1976) 'Remarks on the Theory of Collective Choice', *Economica* 43 (172): 381–90.

Tadros, Victor (2019) 'Localized Restricted Aggregation', in David Sobel, Peter Vallentyne, and Steven Wall, eds., *Oxford Studies in Political Philosophy Volume 5*, pp. 171–203, Oxford: Oxford University Press.

Thoma, Johanna (2020) 'Instrumental Rationality without Separability', *Erkenntnis* 85 (5): 1219–40.

Tucker, A. W. (1980) 'A Two-Person Dilemma', *The UMAP Journal* 1 (1): 101.

Tullock, Gordon (1964) 'The Irrationality of Intransitivity', *Oxford Economic Papers* 16 (3): 401–6.

Tversky, Amos (1969) 'Intransitivity of Preferences', *Psychological Review* 76 (1): 31–48.

(1975) 'A Critique of Expected Utility Theory: Descriptive and Normative Considerations', *Erkenntnis* 9 (2): 163–73.

Ullmann-Margalit, Edna and Sidney Morgenbesser (1977) 'Picking and Choosing', *Social Research* 44 (4): 757–85.

Uzawa, Hirofumi (1956) 'Note on Preference and Axioms of Choice', *Annals of the Institute of Statistical Mathematics* 8 (1): 35–40.

(1960) 'Preference and Rational Choice in the Theory of Consumption', in Kenneth J. Arrow, ed., *Mathematical Methods in the Social Sciences, 1959*, pp. 129–48, Stanford, CA: Stanford University Press.

van Inwagen, Peter (1983) *An Essay on Free Will*, Oxford: Clarendon Press.

von Neumann, John and Oskar Morgenstern (1944) *Theory of Games and Economic Behavior*, Princeton: Princeton University Press.

(1947) *Theory of Games and Economic Behavior*, Princeton: Princeton University Press, 2nd ed.

von Wright, Georg H. (1951a) 'Deontic Logic', *Mind* 60 (237): 1–15.

(1951b) *An Essay in Modal Logic*, Amsterdam: North-Holland.

Wakker, Peter P. (2010) *Prospect Theory: For Risk and Ambiguity*, Cambridge: Cambridge University Press.

Williams, Bernard (1973) *Problems of the Self: Philosophical Papers 1956–1972*, Cambridge: Cambridge University Press.

(1976) 'Moral Luck', *Proceedings of the Aristotelian Society, Supplementary Volumes* 50 (1): 115–35.

Acknowledgements

In writing this work, I have been helped, greatly, by a large number of people. I wish to thank Arif Ahmed, Gustav Alexandrie, Paul Anand, Gustaf Arrhenius, Ralf M. Bader, Ken Binmore, John Broome, Krister Bykvist, Timothy Campbell, John Cantwell, Richard Yetter Chappell, Adam Elga, Tomi Francis, John Halstead, Peter J. Hammond, Sven Ove Hansson, Anders Herlitz, Karim Jebari, Petra Kosonen, Kacper Kowalczyk, Kevin Kuruc, Jake Nebel, Martin Peterson, Wlodek Rabinowicz, Daniel Ramöller, Gerard Rothfus, Joe Roussos, Katie Steele, H. Orri Stefánsson, Dean Spears, Johanna Thoma, Fredrik Viklund, Peter P. Wakker, the audiences at Foundations of Normative Decision Theory, 21 June 2018 at University of Oxford, Foundations of Utility and Risk 2018, 26 June 2018 at University of York, and The Stockholm Region Workshop on Economics and Philosophy, 6 June 2019 at Institute for Futures Studies, Stockholm, and two anonymous reviewers for valuable comments. Wlodek Rabinowicz also co-wrote the paper 'A Simpler, More Compelling Money Pump with Foresight', *The Journal of Philosophy* 117 (10): 578–89, 2020, which covers a lot of the same material as Section 2.1. Section 5 contains some material from the paper 'The Sequential Dominance Argument for the Independence Axiom of Expected Utility Theory', *Philosophy and Phenomenological Research* 103 (1): 21–39, 2021. Financial support from the Swedish Foundation for Humanities and Social Sciences is gratefully acknowledged.

Cambridge Elements ≡

Decision Theory and Philosophy

Martin Peterson
Texas A&M University

Martin Peterson is Professor of Philosophy and Sue and Harry E. Bovay Professor of the History and Ethics of Professional Engineering at Texas A&M University. He is the author of four books and one edited collection, as well as many articles on decision theory, ethics and philosophy of science.

About the Series

This Cambridge Elements series offers an extensive overview of decision theory in its many and varied forms. Distinguished authors provide an up-to-date summary of the results of current research in their fields and give their own take on what they believe are the most significant debates influencing research, drawing original conclusions.

Cambridge Elements ≡

Decision Theory and Philosophy

Elements in the Series

Printed in the United States
by Baker & Taylor Publisher Services